Changing Life

Cultural Politics

A series from the Social Text Collective

Aimed at a broad interdisciplinary audience, these volumes seek to intervene in debates about the political direction of current theory and practice by combining contemporary analysis with a more traditional sense of historical and socioeconomic evaluation.

13. *Changing Life: Genomes, Ecologies, Bodies, Commodities*
    Peter J. Taylor, Saul E. Halfon, and Paul N. Edwards, editors
12. *Will Teach for Food: Academic Labor in Crisis*
    Cary Nelson, editor
11. *Dangerous Liaisons: Gender, Nation, and Postcolonial Perspectives*
    Anne McClintock, Aamir Mufti, and Ella Shohat, editors
10. *Spectacles of Realism: Gender, Body, Genre*
    Margaret Cohen and Christopher Prendergast, editors
9. *Cultural Materialism: On Raymond Williams*
    Christopher Prendergast, editor
8. *Socialist Ensembles: Theater and State in Cuba and Nicaragua*
    Randy Martin
7. *The Administration of Aesthetics: Censorship, Political Criticism, and the Public Sphere*
    Richard Burt, editor
6. *Fear of a Queer Planet: Queer Politics and Social Theory*
    Michael Warner, editor
5. *The Phantom Public Sphere*
    Bruce Robbins, editor
4. *On Edge: The Crisis of Contemporary Latin American Culture*
    George Yúdice, Jean Franco, and Juan Flores, editors
3. *Technoculture*
    Constance Penley and Andrew Ross, editors
2. *Intellectuals: Aesthetics, Politics, Academics*
    Bruce Robbins, editor
1. *Universal Abandon? The Politics of Postmodernism*
    Andrew Ross, editor

Peter J. Taylor
Saul E. Halfon
Paul N. Edwards
editors

Changing Life

Genomes
Ecologies
Bodies
Commodities

Cultural Politics / Volume 13
University of Minnesota Press
Minneapolis ▸ London

Earlier versions of the following essays appeared in *Social Text* 42 (volume 13, number 1, spring 1995): *Do Androids Pulverize Tiger Bones to Use as Aphrodisiacs?* by Simon A. Cole; *Genetic Engineering, Discourses of Deficiency, and the New Politics of Population* by Herbert Gottweis; *Discipline or Solidarity? Ecology as Politics* by Yrjö Haila (*Social Text* version coauthored with Lassi Heininen). Copyright 1995, Duke University Press. Reprinted with permission. *The Terminator Meets Commander Data: Cyborg Identity in the New World Order* by Paul N. Edwards is based on materials drawn from *The Closed World: Computers and the Politics of Discourse in Cold War America* (Cambridge: MIT Press, 1996), used by permission. *Bubbles in the Cosmic Saucepan* by Rosaleen Love originally appeared in *Arena Magazine* (Australia) no. 5 (June–July 1993). *Arena Magazine* is available from P.O. Box 18 North Carlton, 3054 Australia, used by permission. *Contradictions along the Commodity Road to Environmental Stabilization: Foresting Gambian Gardens* by Richard A. Schroeder originally appeared in *Antipode* 27, no. 4 (1995), used by permission of Blackwell Publishers. The first half of *How Do We Know We Have Global Environmental Problems? Undifferentiated Science-Politics and Its Potential Reconstruction* by Peter J. Taylor is a revised version of an article coauthored with F. Buttel, reprinted from *Geoforum* 23, no. 3, 405–16, copyright (1992) with kind permission from Elsevier Science Ltd, The Boulevard, Langford Lane, Kidlington OX5 1GB, UK.

Published by the University of Minnesota Press
111 Third Avenue South, Suite 290
Minneapolis, MN 55401-2520

http: //www.upress.umn.edu

Printed in the United States of America on acid-free paper

Library of Congress Cataloging-in-Publication Data

Changing life : genomes, ecologies, bodies, commodities / Peter J. Taylor, Saul E. Halfon, and Paul N. Edwards, editors.
p. cm. — (Cultural politics ; v. 13)
Includes index.
ISBN 0-8166-3012-7 (alk. paper). — ISBN 0-8166-3013-5 (pbk. : alk. paper)
1. Life sciences—Social aspects. 2. Life sciences—Political aspects.
I. Taylor, Peter J., 1955– . II. Halfon, Saul E. III. Edwards, Paul N.
IV. Series: Cultural politics (Minneapolis, Minn.) ; v. 13.
QH333.C48 1997
570'.1—dc21 97-14055

The University of Minnesota is an equal-opportunity educator and employer.

Contents

Introduction: Changing Life in the New World Dis/Order

Paul N. Edwards, Peter J. Taylor, and Saul E. Halfon

In a more Puritan vein, my scopophilic curiosity about and frank pleasure in the recent doings of flounders and tomatoes must not distract attention from what is entailed by such new kinship relations in the conjoined realms of nature and culture. Large commercial stakes, with attendant national and international intellectual property issues, are involved. Hunger, well-being, and many kinds of self-determination—implicated in contending agricultural ways of life with very different gender, class, racial, and regional implications—are very much at stake. Like all technoscientific facts, laws, and objects, seeds only travel with their apparatus of production and sustenance. The apparatus includes genetic manipulations, biological theories, seed genome testing practices, credit systems, cultivation requirements, labor practices, marketing characteristics, legal networks of ownership, and much else.
▸ *Donna Haraway*[1]

In uneven and sometimes contradictory ways, new concepts of life and new life-forms—human, nonhuman, and artificial—are now implicated in the reshaping of social order and disorder. The changing cultural-political economy of life links the categories,[2] texts, test tubes, databases, and simulations of science to diverse processes: to the ever-expanding and ever more rapid circuits of information, finance, and commodities; to the declining regulatory state as it makes space for these ascendant transnational networks; and to capital's extension of its legal domain over intellectual property, life-form patents, and marketable pollution licenses. Changes in life have also evoked both resistance and participation by "new" social movements. In their discourses, globalized responsibility for sustaining the environment coexists with the promotion of individualized responsibility for disease and health. And, while some peoples fear being pushed further to the margins through the production of new hybrids, others give a liberatory spin to their visions of more extensive coupling with machines.

Changing Life studies the changing sciences and technologies (S&Ts) of life (from planetary management to genomic sequencing) and examines the life (from ecologies to cyborgs) being changed. The essays address transformations at the center of transnational relations and at the margins; during the past, post-Cold War decade and on the longer wave; in fantasies and in actuality. Our contributors, drawn from a range of disciplinary persuasions within science and technology studies (STS),[3] and also from geography, ecology, and developmental biology, provide a range of interpretative angles on the metaphors, narratives, models, and practices of life sciences. We argue that one cannot understand the power of the life sciences in modern society without simultaneously probing the intersections of cultures of S&T with other cultural arenas. The collection of essays thus represents a collective attempt to join the insights of the two emerging interdisciplinary fields of STS and cultural studies.[4] Finally, as a work of cultural *politics*, this volume aims to make its own contribution to changing life in a context of changing social dis/order.

In different ways, all of the inquiries that led to the essays of *Changing Life* arose in response to the New World Order, a cultural-political-economic construction of globality in the post-Cold War era. After the fall of the Berlin Wall in 1989, Western capitalism presented itself as having achieved its grandest goal: a global market economy joining all the world's nations. The dominating influence both of capitalist hegemons like the United States and of the "external world" of closed, autarkic Communist societies had been swept away in the unifying tide of a new, trade-based, multipolar system. Following the model of the 1991 Persian Gulf War, the United States would henceforth lead by example rather than by sheer force, with United Nations-sponsored international collaborations replacing unilateral U.S. interventions abroad. Genuine world peace, or at least a world peacekeeping force capable of dampening and containing regional conflict, seemed poised to rise from the ashes of history's most expensive and dangerous arms race.

The new globalism has many dimensions beyond trade and military conflict, and each aspect weaves in developments in S&T. In the domain of the environment, for example, the Montreal Protocol on the ozone layer was dramatically strengthened in 1989 (as the antarctic ozone hole spread northward toward Australia) with a near-total phaseout of ozone-depleting chlorofluorocarbons planned by the year 2000. In 1992, the huge United Nations Conference on Environment and Development in Rio de Janeiro produced a series of voluntary international agreements on biodiversity, deforestation, and Third World development, as well as a Framework Convention on Climate Change that laid groundwork for eventual binding agreements on greenhouse gas emissions. Preparing for these agreements, hybrid science-policy organizations, such as the Intergovernmental Panel on Climate Change, worked not only to build agreement on scientific issues but simultaneously to bring a range of political players into the consensus-building process.

Science-based international environmental policy and UN-led peacekeeping would together be the leading edge of a new global culture of consensus.[5]

The Human Genome Project, building on the universalism of the genetic code and refining its techniques in pilot projects on organisms no longer so far from humans as E. coli, yeast, and nematodes, also promised an S&T product for the whole human species. By the early 1990s human genetic maps and sequence data from various international collaborations—and from vigorous competition among parallel projects—were being assembled in computer databases. The genomes of diseased families could be compared against the normal ones and markers for specific diseases identified. Reproductive counseling and health advice could be readily given; therapies specifically tailored to a person's genetic profile would soon follow. Meanwhile, the power of information-based approaches to biology and medicine was invoked by computer scientists at the Santa Fe Institute for Complexity Studies (and elsewhere) when they started to create "artificial life," or A-life: simulated life-forms, complete with their own "genetics," "sex," and "reproduction," dwelling and "living" inside the circuits of silicon-based machines.

Finally, new computing and communications technologies promised vast social and political changes, moving toward a new and more interactive version of McLuhan's "global village." Students with fax machines brought the Communist Chinese propaganda machine to its knees during the Tiananmen Square uprising of 1988, discrediting the sanitized official accounts of the massacre. The high-tech U.S.-led forces of the Persian Gulf War flattened the Iraqi army with cruise missiles, pinpoint Global Positioning System real-time surveillance, and laser-guided bombs. Electronic mail, which connected more than 140 nations by 1994, became a vast information funnel for unofficial news during the Balkan civil war and many other conflicts; connections to the Internet increased at geometric rates. Beginning in 1994, the World Wide Web brought graphic interfaces and a vast variety of information-retrieval tools to the desktop of anyone with a PC, a modem, and an Internet connection. To many observers, these media marked the long-awaited infrastructure of an emerging global network culture that would transform the world.

When the smoke cleared from the oil fires of Kuwait and the wreckage of Euro-Soviet Communism, the New World Order revealed itself to be anything but orderly, new, or worldwide. Ethnic and intranational ("tribal") warfare erupted throughout the former Communist states and across the African continent, defying UN and NATO peacekeeping efforts, which sometimes even exacerbated the problem. Close inspection of international environmental regimes revealed rampant noncompliance, fragmented consensus, unenforced (and unenforceable) voluntary agreements—and an emerging black market in chlorofluorocarbons. "Global" computer networks began their slide from public good (albeit a "public" limited to those hooked into the technical infrastructure) to one more medium for private corporations to shape content and access around advertising targeted on particular, elite demographic segments. Despite the hosannas of economists and cyberpundits, the newly extended market economy and the global information instrastructure brought with them nothing remotely resembling a free exchange among different cultures. Looking back on the advent of fully computerized war revealed that the Patriot antimissile missile had failed to bring down even a single Iraqi Scud, and instead scattered its own debris across Israeli cities to lethal effect.[6]

The New World Order was oversold; that is now apparent to all. Critics who had quickly identified its hype and anticipated problems in its realization should, however, take little comfort in this outcome. In a sense we always faced a New World Dis/Order—all constructions of order can be interpreted as expressing the disorders that are most feared.[7] This remains the case even now that the dominant rhetoric focuses on certain disorders,[8] and the current processes of cultural-political-economic construction are just as formidable as the confident triumphalism that preceded them.[9] It is in this context, then, of a post-Cold War world dis/order that *Changing Life* makes its contribution to critical inquiry about science, technology, and culture.

Changing Life also arises in the context of developments in intellectual and scholarly culture. The two transdisciplinary fields to which it contributes—cultural studies and STS—are themselves dis/orderly; the project of joining them is not without its troubles.

Cultural studies may be characterized as applying the insights of cultural anthropology, critical theory, literary studies, and other fields to explore the politics of contemporary cultures. As a loosely defined movement, it has mounted a revolt against traditional framings of politics, turning from government, economics, and military force toward culture and civil society as the critical arenas where political power is created, renewed, and exercised. Its key technique—descended by many separate avenues from Gramsci's concept of the "subaltern"—has been the resurrection of alternatives to the monolithic liberal "we." Cultural studies often attempts to amplify, to validate, and thus to reinscribe the muffled voices of invisible and oppressed groups in all their myriad complexity. To complement and justify this reinscription, cultural studies has sought simultaneously to establish the situatedness of purportedly universal or totalizing accounts and to expose the privilege such accounts afford to dominant processes and groups.

For these goals and purposes (as well as for others), the practices and knowledges of S&T are exceptionally fertile ground in which to work. Science has a history of making universal, contextless claims; technological "advances" are equated with progress; all the while, S&T simultaneously serve powerful institutions, such as the military and the megacorporation. The products of S&T—technical facts and technological artifacts—when they enter cultural life beyond the laboratory and the factory, as they transform it and are transformed by it in turn, have often proven far more ambiguous, unstable, and dangerous than their developers and promoters imagined. Users in situ have had, in different cases, both more and less power over the uses and interpretations of science and technology than those envisioned for them by scientists, engineers, and the institutions within which they work.[10]

Cultural studies of S&T have generated a lot of heat among, on one hand, scientists who see the authority of scientific knowledge under challenge and, on the other, cultural and STS interpreters of science. Similarly, technological optimists have sparred with those having a darker or more ambivalent view of technological change.[11] This friction has exacerbated divisions within cultural studies and, even more so, within STS.[12] The contributors to this volume, however, do not wrangle over the desirability and correct formulation either of these fields or of their fusion. Instead, *Changing Life* addresses the conservatism evident in the attacks on (and within) STS and cultural studies by pushing the cultural studies of S&T project forward.[13] In this spirit:

- We assert the importance of S&T in general as a subject for cultural studies, and, especially for us, the involvement of S&T in reconstructions, both material and cultural, of living processes and social life.
- We transgress boundaries, in many senses. Our collection is unashamedly transdisciplinary, and aims to subvert the hierarchical intellectual division of labor among the humanities, social sciences, and natural sciences. Several of the authors are scientists by profession or training; their inclusion heralds the potential for further transdisciplinary work involving not only commentators on or interpreters of science, but also scientific practitioners themselves. The collection's range of writing style breaks from the still-dominant positivistic conventions of the social and natural sciences; it demonstrates the space for play (and for work) that can be opened up, thus extending the range of possibility for social, cultural, and political studies. And the processes discussed emerge in a context that spans

levels—from the local to the transnational, and from genomic to planetary management—and destabilize conventional or "natural" boundaries. Indeed, central to several essays are boundary transgressors such as cyborgs and androids.

- We embrace the politically transformative dimension of scholarly inquiry. The different political layers of *Changing Life* include an insistence on transdisciplinarity at a time when this is under covert and overt attack in the academy; promoting an expansion of the interpretative resources employed by critics of the dominant cultural-political economies that are based around the new life sciences and technologies; and opposing ahistorical idealizations of local knowledge, simplistic imperatives of ecological stabilization (especially at the global level), and restrictive bodily and sexual discourses.
- We acknowledge the necessary partiality of *Changing Life*. The contributors seek to join with others in changing life in a changing social dis/order. We hope that, through the many and diverse resources this collection provides, we are contributing to diverse interventions into processes linking genomes, ecologies, bodies, and commodities.

Paul N. Edwards, a historian of technology and American politics, uses shifts in images of cyborgs, or human-technology hybrids, as a barometer of cultural change. Edwards explores three salient elements of early 1980s political and cultural discourse: the Cold War, mass-culture computerization, and artificial intelligence. Across the cusp of the Cold War's end, each of these elements saw significant, and related, change. The Cold War, focused on the superpowers and the ever-present threat of nuclear holocaust, gave way to the lopsided, "clean" Persian Gulf War. Computers moved from the military and corporate world into the mass culture of consumer electronics, changing (for many) from ill-understood foe to commonplace friend in the process. Paradigms in artificial intelligence graduated from the early 1980s promise of expert systems, with their rationalistic approaches that sought to spell out their context of action in advance using abstract rules and facts, to neural networks, which acquired contexts of action by learning, virtually without programming or precalculated strategies.

Via a close reading of two science-fiction films, *The Terminator* (1984) and its sequel, *Terminator 2: Judgment Day* (1991), Edwards argues that, taken together, these transitions have dramatically altered the grounds for Donna Haraway's cyborg politics. *The Terminator*, with its dark postholocaust future in which humans fight a losing battle against intelligent computerized machines, mirrored the Reagan Cold War of the 1980s. The grim exigencies of humanity's constant war with its own creations shape both the future and the present in the image of emotionless military discipline. This imagery reflected the politics of what Edwards calls the "closed world" of the Cold War, a world divided against itself in a permanent state of self-destructive tension—one within which cyborg politics, a politics of taking in hand the assembly of new cultural identities from the fragmented remains of decaying dualisms, seemed the only genuine option. *Terminator 2*, released at the cultural moment of the Persian Gulf War and the New World Order, captured the transition to a very different world. In this film the cyborg is returned to its place as technological servant and gender roles challenged in the earlier film are replaced with altered, but still recognizably traditional, versions of femininity and masculinity. An enemy or a cultural Other during the Cold War, cyborgs thus became a friend and merely other in the cultural pluralism of the New World Order.

Changing images of bioorgs, or biological organisms, as they are still called, are also culturally revealing. Scott F. Gilbert, a developmental biologist, delineates four post-World War II representations of the physical body, each of which, he argues, is both literally and metaphorically paralleled by a different view of the nature of the self—and also by a different view of the relationship between science, culture, and truth in the form of the "body politic." The "neural body" is the thinking body, the Cartesian self, with the brain as both controller and center of identity, at the top of a hierarchy where abstraction reigns supreme over the mundane particularity of physical reality. In the parallel "neural view" of science, science functions as a kind of disembodied thought: it locates acultural truths, independent of material circumstances. The "immune body" is a defensive structure, built around the distinction between Self and Nonself; its modern paradigmatic diseases, with numerous political parallels, are no longer infection—an invasion metaphor—but cancer and Alzheimer's: metaphors of an enemy within. The parallel "immunological view" of science sees it as existing to construct, maintain, and defend a particular, Western, male-dominant culture. The "genetic body," like the neural and immune bodies, is concerned with information rather than materiality: it is the body as product of genes, identity as determined by information-bearing molecules. The genetic view of science, correspondingly, sees science as not only independent but *determinant* of culture; it has become society's master controlling element by means of the technology it allows (forces?) us to produce. Gilbert himself wishes to promote a fourth view of body, self, and science. This fourth perspective regards the body as phenotype. The phenotypic body combines the neural, immune, and genetic bodies. But instead of focusing on their abstract information content, it privileges their materiality. It is particularistic, physical, an "integrated network of gene and flesh, brain and gonad, inside and outside." The "phenotypic view" of science, similarly, sees the material products and practical utility of science as more important than its abstract ideas. It places science squarely within a social world where real consequences matter more than theories.

Gilbert builds on his analysis of representations of the body to advance a novel approach to thinking about the goals of and reasons for including science in multicultural university curricula. He employs the phenotypic view to argue for a new role for science in liberal education. Since curricula are inherently political—and he briefly recounts the politicized history of "Western Civilization" courses to demonstrate the point—Gilbert proposes that biology should replace physics and mathematics as the central science. Biology deals directly with social and political issues—race, gender, health, food, birth control, ecology—and it is fast becoming one of the most significant sciences for industry as well. Finally, now more than ever, biology provides models for society: models for the body politic, metaphors of origin, and defining narratives for modern culture.

Herbert Gottweis amplifies Gilbert's claim that biology is becoming central to cultural-political construction processes in the West. Gottweis, a political scientist, presents genetic engineering as a tool to produce identities in the "overflowing political" arenas of trade wars, regional development, electoral campaigns, and new social movements. Like many of the essays in *Changing Life*, this chapter argues that genetic research must be understood as simultaneously scientific, social, and political in nature. In an intricate intertwining of strands from science, culture, economics, and the state, Gottweis examines the evolution of genetic science after its adoption by the Rockefeller Foundation in the mid-1930s. With the emergence of strong state-funded research after World War II (via the National Institutes of Health and other agencies), this financial and institutional boost produced "molecular biology"—oriented scientifically toward the discovery and analysis of genetic macromolecules, while socially and politically contextualized in terms of benefits to human health and agriculture (at a time when the far more blatantly interventionist post-World War I eugenics had fallen into decline and disrepute). Molecular biology offered a subtle new form of (state-sponsored) interventionism, and the discourse that surrounded it began to include imagery of social "challenges and threats" that the new science could promise to resolve.

In the 1970s, molecular biology spawned "genetic engineering." These new techniques shifted the central focus from genetic molecules to manipulable genetic sequences. As postwar U.S. hegemony declined in favor of a new, far more global economy, genetic engineering generated a new "discourse of deficiency" in which genetic technologies might "improve" upon natural germ lines in virtually any species or kind of living organism, from bacteria to plants to human beings. Gottweis argues that this shift fit neatly with other emerging technologies and economic structures to produce a new form of "technopanopticism"—and rewrote subjectivity in the process, as it transformed and normalized dominant views of "human nature." Finally, Gottweis explores the varieties of social resistance to the genetic "discourse of deficiency."

The "overflowing political" arenas discussed by Gottweis for genetic engineering extend also to ecology and to the political peripheries. Social geographer Richard A. Schroeder analyzes a Gambian agroforestry initiative in the 1980s and its market-based strategy for ecological stabilization, a "commodity road" for ecological stabilization increasingly popular in international development circles. After the initial attempts to establish village woodlots failed, the emphasis shifted to fruit trees, an initiative that engendered a new political alliance of the state, nongovernmental organizations (NGOs), and male landholders against women gardeners. The women's market gardens—an outcome of earlier development projects aimed to help women—are being converted into men's fruit tree orchards. The conflicts generated within the peasant households are simultaneously struggles over meaning and struggles over land, labor, and production. Dissent and consent are manufactured in the pursuit of ecological-economic goals.

Yrjö Haila, an ecologist, explores further the theme of ecology as politics. Haila argues that all conceptions of ecological problems carry with them not only images of the natural order, but ideas about the social order as well. Haila shows that a principle he names "social order first" governs most ecological thinking. In other words, most treatments of ecological problems assume that the human order should and will remain much the same, no matter what the problem faced or the solution proposed. Using as examples the widespread conception of the Chernobyl nuclear accident as an environmental crisis, on the one hand, and recent attempts to develop notions of "environmental security" in parallel with the logic of military or national security, on the other, he holds that the "social order first" principle constitutes a "disciplinary logic" that severely limits the possible range of public discourse. Haila distinguishes three main modes through which assumptions about the governability of society have been incorporated into ecological programs. In the "naturalization" mode, frictions in humanity-nature interactions are assumed to arise from natural human features such as population growth and appear, consequently, unsolvable. In "linearization," problems are taken to be most efficiently resolved with a step-by-step approach and trust in technological innovation. In "moralization," the solution is held to lie in a new spiritual relationship with nature. In these ways and in others, Haila argues, ecology has become a new discipline for disciplining.

The preeminent discourse of disciplining in the name of the environment is that surrounding the so-called population problem. Saul E. Halfon, who studies the politics of science and technology, explores the content of recent United States policy making concerning "overpopulation." He traces the production and continual reconstruction of overpopulation discourse since World War II. At its multiple and sometimes contradictory cultural sites, overpopulation can be linked to environmental imperatives, to shifting development goals for the Third World, and to the production of gendered, racialized, and medicalized bodies. Such a map of knowledge and power opens up considerably the cultural politics around overpopulation.

Neo-Malthusian environmentalism was a forerunner to and remains a prominent strand of the globalization of environmental science and politics. Peter J. Taylor, who works at the intersection of science studies and environmental studies, interprets this discourse of globalization as privileging moral and technocratic views of social action. Discounted are the difficult politics that stem from differentiated social groups within any place or region having different interests in causing and alleviating environmental problems. In a concluding reflection, however, he acknowledges that this broad-brush interpretation, like the frameworks critiqued, attempts to cut through the unequal and heterogeneous practical and conceptual facilitations of science and political mobilization. Attempting to make the politics of cultural interpretation more explicit, he formulates his interpretative propositions as useful heuristics that can be employed to expose more of the diverse practices of planetary scientists and environmental activists.

Simon A. Cole, an anthropologist of technics, comments on the many ironies that surround the preservation of the genetic lines of endangered species by commercializing them. Cole offers a response to the question begged by much contemporary discourse on the extinction of "charismatic megafauna": Why do we care? What emotional and cultural chords are struck by the demise of other species? To explore the cultural moment of the "tragedy of extinction," Cole introduces the concept of "animal empathy," which he takes from the 1968 novel *Do Androids Dream of Electric Sheep?* by Philip K. Dick. Modern postindustrial culture hauntingly resembles Dick's fictional world where, in the wake of a nuclear holocaust that has destroyed most wild animals, the love of animals has literally become a religion and pet ownership has become a major symbol of social status. In Dick's novel, the capacity for animal empathy also marks human uniqueness in relation to the android "replicants" they have created. Weaving together a discussion of Dick's novel, the 1982 film *Blade Runner* that it inspired, and the "culture of extinction" as generated variously by environmentalists, conservation biologists, and other elements of postindustrial Western societies, Cole evokes the many paradoxes of the "tragedy of extinction." Noting that Western environmentalists have targeted the Asian uses of rare animals as medicines and aphrodisiacs, he points out that Western discourse on extinction often cites the loss of genetic information potentially usable in pharmaceuticals as a major reason to deplore and prevent the extinction of species (at least before their genetic codes can be inventoried); that is, the use of endangered species for health-related purposes is not only acceptable but crucial when fit within the context of Western biotechnology—yet when the same uses take place within the context of Asian conceptions of medicine and health, they are demonized. In what Cole names "blade-runner society," the politics of endangered species sometimes "embrace[s] the very values that led to endangerment in the first place."

Rosaleen Love, who teaches science communication and writes science fiction, takes flight from the fact that so much of the scientific information with which we are daily overwhelmed is conflicting and contradictory. Her essay blends aspects of the public perception of science into a fantasy/meditation on possible interrelations between the greenhouse effect, chaos theory, Gaia, the anthropic cosmological principle, and electromagnetic pollution of the environment.

Changing Life aims to enlarge the community of participants in both cultural studies and STS and to add to the emerging links between these two areas of scholarship. In an afterword, Peter J. Taylor identifies problems calling for further work in both STS and cultural studies. He starts from sociology of scientific knowledge (SSK), which has undermined philosophers' accounts of how scientists establish knowledge by examining how scientists as actual, not idealized, agents make their science. Such questions of epistemology and *agency*, however, become more complex than those SSK has tackled once one acknowledges, as *Changing Life* certainly does, the large and heterogeneous arena in which life is (re)constructed, an arena extending from genetically hybrid organisms to transnational economies. To better address such questions and to engage in these (re)constructions, five "shifts in positioning" are needed in cultural studies/politics of S&T and in STS. Through the first shift scientists come to be treated as agents who construct jointly their knowing and intervening by mobilizing heterogeneous resources; the other four shifts enlarge and enhance this picture. The point, however, is not to refine our accounts of how scientists work in different contexts. Instead, Taylor argues that although the five shifts are already under way in particular sectors of cultural studies and STS, the shifts need to be pushed further and *applied to a wider class of agents, ourselves included, who not only interpret, but also intervene in science, technology, and culture*.

Notes

1. Donna J. Haraway, "Mice into Wormholes: A Technoscience Fugue in Two Parts," in *Cyborgs and Citadels: Anthropological Interventions on the Borderlands of Technoscience*, ed. Gary Downey, Joseph Dumit, and Sharon Traweek (Seattle: University of Washington Press, 1997).
2. The historical constitution and ongoing reconstitution of "life" as a category need to be addressed more critically than a book title and the style of this Introduction allow. See Stefan Helmreich, "Replicating Reproduction in Artificial Life: Or, the Essence of Life in the Age of Virtual Electronic Reproduction," in *Reproducing Reproduction*, ed. Sarah Franklin and Helena Ragoné (Philadelphia: University of Pennsylvania Press, 1996).
3. We have mostly used the terms "science and technology studies" (STS), "science and technology" (S&T), and "scientist" for practitioners of S&T. Other writers have used, more or less equivalently, the terms "science studies," "technoscience" (to describe science, engineering, technology, and their variously interwoven combinations), and "technoscientist."
4. Other recent anthologies linking cultural studies and STS include Stanley Aronowitz, Barbara Martinsons, and Michael Menser, eds., *Technoscience and Cyberculture* (New York: Routledge, 1995); Chris H. Gray, ed., *The Cyborg Handbook* (New York: Routledge, 1995); and Downey, Dumit, and Traweek, eds., *Cyborgs and Citadels*. By quoting Donna Haraway to start this Introduction and by another long quote from Sharon Traweek in the Afterword, we acknowledge their leadership role in developing the exchange between STS and cultural studies. See Donna J. Haraway, *Primate Visions: Gender, Race, and Nature in the World of Modern Science* (New York: Routledge, 1989), "The Promises of Monsters: A Regenerative Politics for Inappropriate/d Others," in *Cultural Studies*, ed. Lawrence Grossberg, Cary Nelson, and Paula A. Treichler (New York: Routledge, 1992), 295-337, and *Modest Witness@Second Millennium: The FemaleMan(c) Meets OncoMouse(tm)* (forthcoming); and Sharon Traweek, *Beamtimes and Lifetimes: The World of High Energy Physicists* (Cambridge: Harvard University Press, 1992), "When Eliza Doolittle Studies 'enry 'iggins," in Aronowitz et al., eds., *Technoscience and Cyberculture*, 37–55, and "Bodies of Evidence: Law and Order, Sexy Machines, and the Erotics of Fieldwork among Physicists," in *Choreographing History*, ed. Susan L. Foster (Bloomington: Indiana University Press, 1994).
5. For an overview of global environmental science and policy making, see Sheila Jasanoff, and Brian Wynne, "Scientific Knowledge and Decision Making," in *State of the Art Report on Climate Change*, ed. Steve Rayner (Richland, Wash.: Batelle—Pacific Northwest Laboratories, 1997).

6. Theodore A. Postol, "Lessons of the Gulf War Experience with Patriot," *International Security* 16, no. 3 (1991–92): 119–71.
7. This interpretative inversion is best illustrated in Donna Haraway's reading of the early twentieth-century expeditions, taxidermy, and exhibitions of the American Museum of Natural History in terms of white Anglo-American men's concerns about the disruptive, dysgenic, and decadent influences on politics, race relations, and culture associated with rising immigration and women's suffrage (Donna J. Haraway, "Teddy Bear Patriarchy: Taxidermy in the Garden of Eden, New York City, 1908–1936," *Social Text* 11 [1984–85): 20–64; reprinted in *Primate Visions*, 26–58).
8. Recall the newspaper headlines and editorials in the United States during the spring of 1996 worrying about the possibility that Russian voters would reject change or reject freedom if they dumped the "democrat" Yeltsin and elected, democratically, a Communist president to replace him.
9. The observations of Raymond Williams in *The Year 2000* (New York: Pantheon, 1983) are still apt. In the face of a progression of crises, he noted, there has been a tendency for "the practical cancellation of detailed, participat[ory], consciously chosen planning" (11). Crises are "simply exposures of existing real relations" in societies, yet, he argued, it has been crisis-managers, who pursue a "politics of temporary advantage," to whom citizens have been abdicating control of their futures.
10. For an entry point into STS, see Andrew Pickering, ed., *Science as Practice and Culture* (Chicago: University of Chicago Press, 1992). For a description of cultural studies of S&T, giving specifics of the origins of various strands within cultural studies, see Sharon Traweek, "Introduction to the Cultural and Social Studies of Sciences and Technologies," *Culture, Medicine, and Psychiatry* 17 (1993): 3–25. See also the Afterword to this volume; David Hess, *Science and Technology in a Multicultural World: The Cultural Politics of Facts and Artifacts* (New York: Columbia University Press, 1995); and Joseph Rouse, "What Are Cultural Studies of Scientific Knowledge?" *Configurations* 1, no. 1 (1992–93): 1–22. For work in this vein, see references in notes 4 and 12.
11. Paul Gross and Norman Leavitt, *Higher Superstition: The Academic Left and Its Quarrels with Science* (Baltimore: Johns Hopkins University Press, 1994). See responses in Andrew Ross, ed., "Science Wars," *Social Text* 46–47 (1996): 1–252.
12. For dismissive responses within STS to the prospect of a cultural studies of S&T, see Harry M. Collins, "Review of Bruno Latour, *We Have Never Been Modern*," *Isis* 85, no. 4 (1994): 672–74; and Peter R. Dear, "Cultural History of Science: An Overview with Reflections," *Science, Technology & Human Values* 20, no. 2 (1995): 150–70.
13. Other recent work in the vein of cultural studies of S&T includes Andrew Ross, *Strange Weather: Culture, Science, and Technology in the Age of Limits* (London: Verso, 1991), and *The Chicago Gangster Theory of Life* (London: Verso, 1994); Mario Biagoli, Roddey Reid, and Sharon Traweek, eds., *Located Knowledges, Special Edition of Configurations* (1994); Emily Martin, *Flexible Bodies: Tracking of Immunity in American Culture from the Days of Polio to the Age of AIDS* (Boston: Beacon Press, 1994); Rayna Rapp, "Risky Business: Genetic Counseling in a Shifting World," in *Articulating Hidden Histories*, ed. Rayna Rapp and Jane Schneider (Berkeley: University of California Press, 1995); George Marcus, ed., *Technoscientific Imaginaries: Conversations, Profiles, and Memoirs* (Chicago: University of Chicago Press, 1995); and Helmreich, "Replicating Reproduction in Artificial Life." See also notes 3 and 7, and works cited in Sharon Traweek, "Introduction to the Cultural and Social Studies of Sciences and Technologies," *Culture, Medicine, and Psychiatry* 17 (1993): 3–25.

The Terminator Meets Commander Data: Cyborg Identity in the New World Order

Paul N. Edwards

In 1985, Donna Haraway drew the attention of cultural studies and STS to the figure of the cyborg as a way of conceiving political identity in what she called "the world of the integrated circuit."[1] Cyborgs are cybernetic organisms, assembled at the interface between technology and biology, both conceived as problems of information, coding, and control. They straddle the borders between fiction and fact, virtual and actual, machine and human and animal. Haraway argued that the cyborg might afford feminists (and others) a chance to reformulate problems of gender identity in terms that might evade capture within the pairings of ancient philosophico-ideological dualisms like mind/body, technology/biology, and culture/nature. Cyborgs, in their contingency and unnaturalness, could roam the margins and interstices of a society integrated on an increasingly huge scale through information and control technologies, perhaps forming new alliances and finding new power in their release from old essentialisms.

Haraway wrote "A Manifesto for Cyborgs" at a time when information technology was making another of its decadal quantum jumps in scale and scope. The early 1980s saw the introduction of personal computers and the rise of Silicon Valley, Japanese electronics, and the microprocessor industry. They were also the height of the Reagan-Bush Cold War, a massive military buildup led by high technology.[2] The scary sci-fi cyberweapons of the Strategic Defense and Strategic Computing Initiatives demonstrated that if the cyborg could be an ally, it could also quite clearly be an enemy—an icon of multinational capitalism and the awesome killing machine built for a global war.

What has become of cyborg identity in the New World Order? With the collapse of Communism throughout Eurasia, the most powerful ideological dualism of our time disappeared into history, revealing the skeleton of terrorist autarkic nationalism it had for so long concealed. In the Persian Gulf War of 1991, the computerized weapons built for the Cold War were finally used to devastating effect, while another kind of information technology—Pentagon media control—filtered the war and its high-tech weapons into the homes of millions through a carefully constructed information funnel.

Around the same time, new issues, theories, and technologies—global warming, chaos theory, connectionism, and neural networks—focused attention on nondeterministic, unpredictable domains whose principles belied the hubristic rationalism of the grand systems theories on which Cold War science, technology, and strategy had hung. How did cyborg figures respond to these events and transformations?

This essay lays out the beginnings of an answer to these questions through a political reading of two films, *The Terminator* (1984) and its 1991 sequel *Terminator 2: Judgment Day*. If my essay is a political reading of fictional dramas, it is also a fictional reading of political dramas, interpreting politics as a dramatic space whose settings, narrative frames, and central characters help to shape outcomes and identities. I will briefly sketch the political context of each film before proceeding to an analysis that illustrates, perhaps better than any purely factual discussion, the transition from cyborg as enemy and cultural Other to cyborg as friend and merely other in the cultural pluralism of the New World Order.

1984

The first half of the 1980s saw the height of what Fred Halliday has called the "Second Cold War."[3] Ronald Reagan was elected president on the turning political tide of a self-proclaimed "Moral Majority" and a perceived American military "weakness" comprised of both a post-Vietnam failure of confidence and a need for more hardware. Labor union power collapsed in the wake of global capital restructuration, led by Japanese competition and a wave of corporate megamergers, leaving the Republican upper middle class in charge of massive tax cuts for the rich.

Reagan's first term in office was marked by the appointment of dubiously qualified ideologues (such as James Watt and Anne Gorsuch Burford) to high administrative posts, a massive resurgence of Cold War rhetoric, and major increases in military spending. During this period the Pentagon attempted to severely restrict trade and scientific communication in broadly defined defense-sensitive areas, including advanced computing. Reagan ordered toy-war skirmishes in Grenada and Libya designed to flex American muscle on the global stage, though his attempt to throw American weight around in the cauldron of the Middle East (the Marine peacekeeping force in Lebanon) ended in a scurrying retreat. Closer to home, Reagan backed the not-so-secret counterrevolutionary war against the Nicaraguan Sandinistas, and the bloody repressions led by death squads in El Salvador and elsewhere throughout Central America.

Reagan's was also the most popular presidency in history.

In 1983 Reagan proposed the Strategic Defense Initiative (SDI, known popularly as "Star Wars"), his plan for a total space-based nuclear missile defense built with superhigh-technology weapons. The "peace shield," as he called it, would constitute a gigantic technological bomb shelter under which Americans could rest secure from the threat of global holocaust. The lesser-known but related Strategic Computing Initiative, a major and highly controversial program in advanced computing and artificial intelligence (AI), was announced by the Defense Advanced Research Projects Agency in the autumn of the same year. Strategic Computing proposed to build a new generation of autonomous and "brilliant" weapons, including robot tanks, an aircraft carrier "battle manager," and a natural-language-speaking "pilot's assistant" for advanced jet aircraft, around massively parallel processors and a form of AI known as "expert systems."[4]

Star Wars preempted the powerful Nuclear Freeze movement that had threatened to steal the thunder from Reagan's Cold War revival. The most advanced computers and computer software ever constructed would be a sine qua non of any such system, and the SDI channeled vast new Pentagon funding into computer research. Revelations of a long (and long-secret) history of computer failures in NORAD nuclear early warning systems, some of them serious, emerged in the early 1980s, sparking public fears of computer-initiated holocaust.[5] Around that time, in opposition to aspects of Star Wars and Strategic Computing, Computer Professionals for Social Responsibility was founded—marking the second organized politicization of computer experts around the question of strategic defenses. (The first had occurred at the time of the quite similar ABM controversy during the Nixon administration.)

At least until the middle of Reagan's second term in office, American anxiety about nuclear war and ideological polarization with Communism reached heights not seen since the 1950s. The much-remarked science-fiction, movie-script quality of Reagan's "leadership," especially in the military arena, only served to heighten his extraordinary popularity. The American public may never have been completely taken in by Reagan's "peace shield" idea, but polls showed that many cheered its vigorous, can-do technological machismo.

The early 1980s also marked two key events in the history of computing: the introduction of low-cost, truly useful personal computers—the Apple II in 1977, the IBM PC in 1981, and the Apple Macintosh in 1984—and the release of the first commercially viable AI software in the form of so-called expert systems.

Personal computers signaled a major change in the social distribution of computer power. The first electronic digital computers, completed just after World War II, were gigantic, enormously expensive number-crunching machines available only to a tiny scientific and military elite. By the late 1950s, computers had become the core of the SAGE continental air defense system and of many other nascent military command-control projects, from the Ballistic Missile Early Warning System to NORAD and the eventual World-Wide Military Command Control System.[6] But they had also been commercialized for accounting and data processing, marketed primarily to government and large businesses such as banks and insurance companies. In this period they were still accessible mainly to a relatively small and specialized group (although their presence made itself widely felt through computer-generated forms and bills and the ubiquitous buck-passing phenomenon in which computers were blamed for frustrating or mysterious mistakes). Gradually, during the 1960s and 1970s, computers became inexpensive and accessible enough to make their way into the offices of even relatively small business and industrial firms. While universities made computers widely available to students and faculty, visions of a "computer-literate" general public remained distant.

Macintoshes and PCs marked the evolution of computers into a "user-friendly" form accessible to consumers for about the price of a used car. By the mid-1980s not only the machines themselves, but direct experience of computer use, were becoming nearly ubiquitous for middle-income Americans. *Time* chose the personal computer as 1982's "man of the year." Apple's famous four-minute television advertisement for the Macintosh—directed by Ridley Scott,[7] broadcast at the height of the 1984 Olympic Games in Los Angeles, and depicting IBM users as lemmings in blue suits—indicated not only the size and character of the mass marketplace to which the new machines were directed, but the broad generality of cultural interest in their explosive spread through affluent society.

Expert systems were more interesting for the hype they received in the business press and the amazing number of start-up firms built around them than for their rather disappointing commercial performance. Although they were claimed to add "intelligence" to computer performance, in practice they functioned primarily as merely another high-level programming language incorporating heuristic rules. Expert systems were and remain significant, however, because they were the first practical products based on research in artificial intelligence (AI), a term coined in 1956 to describe methods of simulating human problem-solving abilities. They brought the notion of AI into everyday parlance and established it as an (apparently) functional technology.[8]

Expert systems are one form of what is known as "symbolic" artificial intelligence. They consist of sets of facts about a given "knowledge domain" such as medical diagnosis or oil prospecting, sets of rules for inferring conclusions from those facts, and an "inference engine" that uses heuristic logic to reason its way to conclusions from a specific set of facts. The systems' successes, such as they are, derive from their focus on very narrow arenas and their highly restricted uses. Typically, they are quite "brittle"; that is, their ability to reason successfully tends to drop off precipitously when they are confronted with borderline cases, and they totally lack common sense.[9] The medical diagnostic system MYCIN, for example, has been known to conclude from his symptoms that a man was pregnant.

Nevertheless, in some situations expert systems performed as well as or better than their human counterparts. This limited success was taken by many to vindicate the symbolic AI research program in effect since 1956—and with it the Enlightenment enterprise of codifying all human knowledge in the form of abstract principles and catalogs of facts. A team headed by Douglas Lenat embarked on the Cyc project (short for Encyclopedia, a direct reference to Diderot's grand plan), an attempt to construct a vast expert-system database capturing all commonsense knowledge in first-order predicate calculus, thus building the foundation for truly intelligent symbolic AI.[10]

It was into this triple context of Cold War, ubiquitous computerization, and artificial intelligence that James Cameron released *The Terminator* in 1984. The film was a relatively low-budget science-fiction horror movie. Its enormous—and unanticipated—popularity may be partly explained by its powerful resonance with the political and social context of that year, with its ringing Orwellian significance.

The Terminator

The Terminator opens in Los Angeles in the year A.D. 2029 amid the rubble and smoke of a nightmarish postholocaust world. We later learn that an all-out nuclear exchange had been initiated by the "Skynet computer built for SAC-NORAD by Cyberdyne Systems. New, powerful, hooked into everything, trusted to run it all. They say it got smart. A new order of intelligence. Then it saw all people as a threat, not just the ones on the other side. It decided our fate in a microsecond: extermination." The few remaining human beings eke out a miserable existence in grimy underground bunkers, emerging only at night to do battle with the robot killing machines that are now the masters of the planet. To finish off the human resistance, the machines send a cyborg (in this case a humanoid robot encased in living flesh) back in time to the preholocaust present, that is, 1984. The Terminator's mission is to find and kill Sarah Connor, mother-to-be of the future resistance leader John Connor. But the resistance is also able to send a soldier, Kyle Reese, back in time to warn and protect Sarah Connor.

The Terminator murders the two other Sarah Connors in the Los Angeles telephone directory, then comes looking for the third. But Reese has already found her and is following her. When the Terminator attacks, he shoots it many times with a shotgun at close range, but this only stops it for the few seconds Reese and Connor need to escape. The basic structure of the plot from this point on is the standard horror-movie script about a scared, helpless woman pursued by an unstoppable monster/man and rescued by a (male) good guy using ever-escalating violence. After many narrow escapes and Kyle's eventual death, Sarah finally destroys the Terminator (now reduced to a robotic skeleton) by crushing it in a metal press inside a deserted automated factory.

Arnold Schwarzenegger plays the Terminator with a kind of terrifying mechanical grace. Completely devoid of emotion, within seconds of his appearance on the screen he kills two young men just to get their clothes. His mechanical nature is repeatedly emphasized through a number of devices. He has a seemingly symbiotic relationship with all kinds of machines: for example, he starts cars by merely sticking his fingers into the wiring. When shot, he sometimes falls, but immediately stands up again and keeps on lumbering forward. We see him dissect his own wounded arm and eye with an X-Acto knife, revealing the electromechanical substrate beneath his human skin. Perhaps most frightening of all, he is able to perfectly mimic any human voice, enabling him to impersonate a police officer and later Sarah's own mother.

What makes the Terminator so alien is not only his mechanical body, but his computerized, programmed mind. At certain points we see through his camera-like eyes: the scene becomes graphic and reddened, like a bit-mapped image viewed through infrared goggles. Computer-like displays of numbers, flashing diagrams, and menus of commands similar to Macintosh software appear superimposed on his field of vision. The Terminator speaks and understands human language, and his reasoning abilities, especially with respect to other machines (and weapons), are clearly formidable. But he is also a totally single-minded, mechanical being. Kyle warns Sarah that the Terminator "can't be bargained with. It can't be reasoned with. It doesn't feel pity, or remorse, or fear. And it absolutely will not stop—*ever*—until you are dead." As portrayed by Schwarzenegger, the Terminator blends images of a perverse, exaggerated masculine ideal—the ultimate unblinking soldier, the bodybuilder who treats his body as a machine—with images of computer control and robotic single-mindedness, complete with an alien subjective reality provided by the Terminator's-eye sequences.

The main theme of the film is the idea of a final, apocalyptic struggle to save humanity, first from computer-initiated nuclear holocaust and second from the threat of self-aware, autonomous machines grown beyond the limits of human control. But a strong subtheme provides an unusual and very contemporary twist.

Sarah Connor begins the film as a very ordinary waitress whose major purpose in life seems to be to try to get a Friday night date. Resentfully, sometimes angrily ("Come on. Do I look like the mother of the future? I mean, am I tough? Organized? I can't even balance my checkbook"), under the relentless pressure of the Terminator's pursuit, she is educated about the threats the future holds and her role as the mother and teacher of the future savior. She bandages Kyle's gunshot wound. Under Kyle's tutelage she learns to make plastic explosive, and listens carefully as he instructs her in the importance of resistance, strength, and fighting spirit. She eventually saves the wounded Reese by dragging him out of the path of an oncoming truck, shouting in his ear, "*On your feet*, soldier!" in a voice that rings with determination. She, not Kyle, is the one who finally destroys the Terminator once and for all, in one of the film's most powerful moments. In the end she is transformed into a tough, purposeful single mother—pregnant by Kyle—packing a forty-four, driving a jeep, and heading off into the sunset and the oncoming storm just as heroically as any cowboy.

The Terminator shows a single mother as a new kind of heroine: the progenitor and trainer of a race of soldiers fighting for humanity against machines. When Sarah asks Kyle what the women of the future are like, he replies tersely, "Good fighters," and in a dream-memory of the future, Kyle's partner on a combat mission against the machines is female. In this portrait women, no longer shrieking and helpless in the face of violence, take up arms and emerge as men's allies and equals in the increasingly dangerous, alien, and militarized world of the future. The subplot of *The Terminator* is about arming women for a new role as soldiers, outside the more traditional contexts of marriage and male protectorship. The message is also that women are the final defense against the apotheosis of high-technology, militaristic masculinity represented by the Terminator—not only because they harbor connections to emotion and love, as in more traditional imagery, but because they are a source of strength, toughness, and endurance: "good soldiers."

The social reality of 1984 held extraordinary resonances with *The Terminator*'s themes. Public anxiety about nuclear weapons, the revelations of epidemic computer failures in NORAD nuclear warning systems, and the

Strategic Defense Initiative created a highly charged context for the theme of computer-initiated nuclear holocaust. A rising tide of robot-based automation in industry, a new wave of computerization in workplaces based on new personal-computer technology, and the Strategic Computing Initiative's controversial proposals for autonomous weapons matched the film's theme of domination by intelligent machines. (Indeed, one of the film's more effective devices is the constant visual reference to machines and computers: robots, cars, toy trucks, televisions, telephones, answering machines, Walkmans, personal computers.)

With respect to gender issues, the film took its cue from two social developments. First, the highest rates of divorce and single motherhood in history grounded the film's search for a new role for single mothers. Second, starting in the mid-1970s women had become increasingly important as soldiers, filling 10 to 13 percent of all U.S. military jobs by 1985, and with serious proposals to increase the ratio to 50 percent in the Air Force (since physical strength is not a factor in high-tech jobs like flying jet fighters, and since women are supposedly able to handle higher G-stresses than men and thus to stay conscious longer during power turns).[11] Thus the film finds its model for the future of womanhood in the armed forces.

The iconography of this film is built around what I call "closed-world discourse." Closed-world discourse names the dramatic frame of Cold War politics. Fictional closed-world dramas are characterized by entrapment within an artificial world where characters confront the limits of rationality. American strategy for the Cold War—global technological oversight and control—both relied on and helped generate an extended set of metaphors, such as containment, closure, and system, and inserted them into the narrative frame of closed-world drama. Computer technology, which allowed the construction of gigantic, integrated, real-time military control systems as well as of complex abstract models of strategy, supported the resulting closed-world discourse in essential ways. Within this framework, the world became a system subject to technological management and control, with all conflicts interpreted through its lens of apocalyptic struggle.

The alternative to a closed world in fiction is not an open world but a "green world," an apolitical dramatic arena constituted by unbounded natural settings, a search for transcendent experience, and the restoration of broken wholeness. Although a kind of green-world politics also exists, it is shadowy and marginal, and has rarely constituted a major force in world politics. Closed-world dramas—both fictional and political—use green worlds as a resource, often offering brief glimpses into green-world spaces as an escape (when there is one) from the totalitarian claustrophobia of the closed world. Space limitations prevent me from providing more than a sketch of these concepts here, but Table 1 displays them in a schematic form.

Table 1 Closed World and Green World

Closed world	Green world
Enclosed artificial spaces: cities, buildings, fortresses, space stations	Raggedly bounded natural settings: forests, meadows, swamps, deserts
Unity of place: all or most action inside enclosures; transfer between locations in sealed vehicles	Flow of action between natural, urban, and other locations; protagonists may enter closed worlds temporarily
Theme: confrontation with limits and destructive elements of rationality, social conventions, political authority, technology	Theme: restore community and cosmic order by surpassing rationality, conventions, authority, technology
Action: invade or escape boundaries of closed world, build or maintain defenses	Action: explore, seek transcendent powers, liberate or destroy closed worlds
Movement: among nested enclosures (vertical)	Movement: among locations (lateral)
Immanence: human, psychological forces—technology, political power, human will, pain, resistance	Transcendence: magical, natural forces—mystical powers, sexuality, animals, other life-forms, spirits, natural cataclysm
Outcome: self-destruction, exorcism of hyperrational and restrictive social norms, or escape to green world	Outcome: unification or reunification of groups or of men and women; celebration; possible destruction or integration of closed world
Struggle: apocalyptic—good versus evil, self versus dangerous/alien Other	Struggle: integrative—comprehending complexity and multiplicity, grasping Other as merely another
Archetype: siege (*Iliad*)	Archetype: quest (*Odyssey*)

The ambiance throughout *The Terminator* is that of closed-world drama. The Terminator's terrifying, mechanical single-mindedness, and the references to the Skynet "defense network computers," are archetypal closed-world images. The Terminator's mind is fixed, inflexible, preprogrammed, though within its limits extremely clever; it is an icon of pure logic. The Skynet system is "hooked into everything," enabling it to become intelligent and initiate a nuclear holocaust from its vantage point of central control. Almost all of the action occurs either indoors, inside vehicles, on streets, or in alleyways, and much of it takes place at night. The setting is a grim, usually dark, urban landscape. Virtually no natural objects or landscapes appear in the film.

Scenes from the world of A.D. 2029 take this imagery to a maximum, with nothing remaining above ground but the rubble and twisted girders of blasted buildings and the charred remains of dead machines. Human dwellings are underground, dirty, furnished with weaponry, canned goods, and the burned-out hulks of television sets, now used as fireplaces. Only two scenes in the film occur in a natural setting: the few hours Sarah and Kyle spend resting in a wooded area (though even here they hide in a semienclosed space under a bridge), and the final scene in which Sarah drives off toward the mountains of Mexico in a jeep. Thus, in a pattern characteristic of closed-world discourse, a briefly glimpsed green world is the final refuge—when there is one—from the coming apocalypse.

Cyborg imagery is also prominent in the film. The Terminator is a liminal figure: a man who is a computerized machine; a living, flesh-and-blood organism whose core is a metallic, manufactured robot; a thinking, reasoning entity with only one purpose.[12] He seems to be alive, but he cannot be killed. He talks, but has no feelings. He can be wounded, but feels no pain. In a flashback (to the future), we learn that the Terminators were created to infiltrate the bunkers of the resistance by impersonating humans. Dogs, however, can sense them. Dogs, of course, are marginal figures of another sort, connecting humans with the animal, the natural, and the wild—links with the green world. The Terminator is also a military unit, like Kyle Reese, but he is a caricature of the military ideal: he follows his built-in orders unquestioningly, perfectly, sleeplessly, and he has no other reason for living.

But Kyle, too, has an intense single-mindedness about him, likewise born of military discipline. He dismisses his gunshot wound with a disdainful "Pain can be controlled." He speaks of an emotionless life in the world of the future, where humans, like the machines they fight, live a permanent garrison lifestyle. The Terminator is the enemy, but he is also the self, the military killing machine Kyle too has become—and that Sarah herself must become in order for humanity to survive. Reese and the Terminator are twisted mirror images caught in a deadly dance: humans have built subjective, intelligent military machines, but are reduced to a militaristic, mechanical, emotionless subjectivity in order to fend off their own products.

The fictional world of *The Terminator* draws our attention to the ways closed-world discourse is historically and conceptually linked with the figure of the cyborg. Just as facts (about military computing, artificial intelligence, nuclear weapons, and powerful machines) give credibility to fiction, so too fictions (visions of centralized remote control; clean, automated war; global oversight; and thinking machines) give credibility and coherence to the disparate elements that comprise these discourses. We cannot understand their significance without understanding these linkages.

1991

Armed only with shovels, crowbars, and their bare hands, thousands of German citizens met at Checkpoint Charlie in 1989 to tear down the Berlin Wall that had marked the division between East and West for almost thirty years. This event was rapidly followed by the collapse of Communist governments throughout Eastern Europe and in the Union of Soviet Socialist Republics. With these events, four decades of Cold War drew to a close, and with them many of the signal features of closed-world politics.

Although the danger of regional nuclear war may in fact have increased with the multiplication of unstable nuclear-armed states, the danger of global nuclear holocaust went—at least for the present—the way of the Soviet Union. With the accession of quasi-democratic governments and Western-style market economies throughout the former Communist world, the Cold War's ideological conflicts also vanished. At the same time, the rise of Japan and the newly industrialized countries of East Asia brought to a close the United States' four decades of dominance in the world economy. Economically exhausted by the Cold War, and especially by the final Reaganaut orgy of deficit military spending, America began the slow process of reformulating its global role for a post-Cold War world.

In August 1990, the army of Saddam Hussein invaded Kuwait. Soon afterward, President George Bush announced the dawn of a New World Order. Bush meant to promote a new international political and financial arrangement whereby military interventions, both to restore and to keep peace, would be politically sponsored by the United Nations, financially sponsored by the wealthy industrialized countries, and ideologically sponsored by—and run from—the United States. America would provide leadership, troops, and military hardware, while other countries would contribute money and token military forces. The formidable arsenal built for global war would now be turned to police work.

That arsenal's full power, in the age of microelectronics and smart weapons, had never been demonstrated. Until the Persian Gulf War, it remained (from a cultural point of view) an uncertain mythology; some suspected the high-tech weapons would fail, as they had in Vietnam, against a large and determined army of human beings. But soon after the war began, Pentagon officials broadcast videotapes of computer-controlled, laser-guided bombs destroying buildings in Baghdad.

In that moment, a worldwide television audience experienced the joining of the subjectivity of the cyborg with the politics of the closed world.[13] As we rode the eye of the bomb to the white flash of impact, in the eerie virtual reality of the TV image, we experienced at once the elation of technological power, the impotence and voyeurism of the passive TV audience, and the blurring of boundaries between the "intelligent" weapon and the political will. The dazzling—and terrifying—power of high-technology warfare displayed in the sound-bite war in the Gulf became an emblem for the glories of America's waning global hegemony.

Through their TV screens, everyone in America followed cruise missiles as they cut communications antennas in half, or circled bunkers in order to enter politely through the side door before obliterating them and their inhabitants. Computer-guided Patriot missiles were claimed to knock Scuds out of the sky (in fact, all of them missed their targets),[14] splattering debris across Israeli cities to American applause and Israeli consternation, and sent the stock of Raytheon and General Dynamics soaring. American audiences saw "smart" computers *embodied* in weapons, proto-Terminators, seeking out targets and destroying them with awesome force and fully hyped "precision."

The largest parade in U.S. history celebrated the return home of the veterans, most of whom had done little more than wait, sweat, and consume tankards of Coca-Cola in the Saudi desert. Those who fought had for the most part done so at night, in the air, using long-distance weapons of near-nuclear destructive force against which the feared, but less well armed and poorly led Iraqi soldiers turned out to be all but helpless. The "mile of death," a swath of bodies and decimated vehicles more than a mile wide and ten miles long along the desert road leading north out of Kuwait City, marked the rout of the Iraqis and the graves of thousands of civilians slaughtered by "precision" computer-guided weapons.

Both technologies—the weapons and the Pentagon's information control—had worked to perfection. The tiny numbers of American casualties and the Pentagon's careful management of the imagery of suffering made the war seem virtually bloodless, a sort of virtual-reality video game. In the New World Order, the vanished fear of global holocaust left behind it the pure elation of high-tech military power.

By the late 1980s, attention in artificial intelligence had moved away from expert systems, with their abstract, predigested knowledge and symbolic reasoning. It now focused instead on new techniques that—like the cybernetics of the 1940s and 1950s—drew their inspiration from biological models.

By simulating some of the functional aspects of neurons and their synaptic connections, so-called neural networks were able to recognize patterns and solve certain kinds of problems *without* explicitly encoded knowledge or procedures. Each simulated neuron performed the simple function of summing inputs (which, like inhibitory and excitatory synapses in the brain, might have either positive or negative value). Based on this sum, which might include a "weight" added by the neuron's particular summation function, the neuron passed on a value of its own to other neurons. These in turn summed their inputs and passed them on until some final output value was generated. In a pattern recognition task, for example, someone writes a sentence on a writing tablet fitted with a dense two-dimensional matrix of pressure sensors. Each sensor communicates the presence or absence of a mark to an "input layer" neuron corresponding to its location on the grid. The input layer sends signals to one or more middle processing layers. These, in turn, send signals to an output layer where the final interpretation of the marks and spaces as letters composing a sentence appears.

The network contains no explicit representations of letter forms. There are no templates to fit, no predefined procedures to follow, and no features to detect—no abstract facts or rules at all. Instead, the network is "trained" with feedback techniques. The correct interpretation of the letters is sent backwards through the network; an algorithm adjusts the weighting functions of the neurons and synaptic connections and another trial is conducted. After just a few trials, well-designed networks can recognize images and forms whose interpretation had eluded feature-recognizing, template-matching, procedure-following symbolic AI for decades.

Unlike their predecessors, the new systems were fundamentally learning machines. Although they were not literally models of brains, the biological metaphor of neural networks contrasted sharply, and deliberately, with the mind-program analogy of symbolic AI. A new school of programming known variously as "connectionism" and "parallel distributed processing" saw these new techniques as the salvation of computerized intelligence. Other approaches in "nonrepresentational AI" also derided their predecessors' attempts to encode all knowledge in advance instead of having it emerge from a system's interaction with the real world.[15]

Also during the 1980s, networks of another sort—global networks that allowed computers to communicate with each other—began to grow at an exponential rate. DARPA's early experimental research network, the ARPANET, became the MILNET military network. Based on ARPANET protocols and a backbone structure built by the National Science Foundation, the Internet—a global network of computer networks—began to emerge in the early 1980s. At this writing, a new network was being connected to the Internet on an average of once every ten minutes; traffic over the World Wide Web was increasing at the almost unbelievable rate of 341,000 percent per year.[16]

The cyberspace of William Gibson's *Neuromancer* thus became an increasingly palpable and common experience.[17] In September 1991, the prestigious journal *Scientific American* dignified the word with a special issue titled "Communications, Computers, and Networks: How to Work, Play, and

Thrive in Cyberspace."[18] The number of commercial networks connected to the Internet finally exceeded, in the early 1990s, that of networks belonging to government, educational, and nonprofit institutions. Senator Al Gore, who contributed an article to *Scientific American*'s cyberspace issue, rode to the vice presidency partly on the strength of his plan for a national "information superhighway."

These changes formed a vastly different backdrop: the New World Order; the seemingly bloodless video-game victory of computerized weapons in the Persian Gulf; the return to biological models in artificial intelligence, with its implicit retreat from the impossible challenge of symbolic AI; and the wide-open consumer commercialism of new computer technology. Against this background, Cameron released the sequel to *The Terminator*, among the most expensive films Hollywood had ever produced.

Terminator 2: Judgment Day

In *Terminator 2: Judgment Day* (1991), we see how cyborg subjectivity is transformed under the "New World Order." In *T2*, as its ad campaign called it, all the icons semiotically restructured by *The Terminator* (for convenience here, *T1*)—the emotional woman, the mechanical man, the white nuclear family—are systematically reconstituted for a post-Cold War, postfeminist, post-postmodern world through the rehabilitation of the war cyborg.

T2 is set in 1994, three years before the date of the computer-initiated nuclear holocaust projected by *T1* and ten years after the original *Terminator*. The plot, in many respects, is the same extended chase, except that this time the object of pursuit is not a woman, but a child: John Connor, Sarah's son, now ten years old. The new villain is a more advanced Terminator model, the T-1000, sent back from the future to attack John as a boy. The original Terminator, or rather another, identical T-800 unit (Schwarzenegger) captured and reprogrammed to protect John Connor, returns in *T2* as the good guy.

Despite its basic similarities, the dramatic atmosphere of *T2* has changed from the powerful closed-world ambiance of *T1*. Although green-world imagery remains rare, many of *T2*'s scenes are shot in full daylight, in domestic and suburban settings rather than urban ones. The pressured sense of claustrophobia *T1* produced so effectively has been traded for an alternation between safety and danger and a series of tasks (Sarah's rescue; the recovery of Sarah's weapons cache; the attempt on Dyson's life; the destruction of the T-800 chip) having the structure more of a journey than of flight within a single contained space.

The T-1000, if it can be called a cyborg at all, is nothing like the T-800 (the original). It is neither computer nor machine, but a shape-shifting blob of "mimetic polyalloy," a mercury-like liquid metal. It can be shot, blasted, even shattered, but its body simply congeals again, re-forming itself. It has no moving parts, no neural networks, no bodily functions. Its specialty is the ability to morph (a computer-graphics term for seamlessly transforming itself) into exact replicas of other people and things—John's foster mother, a hospital guard, a tiled floor. (In a perfect postmodern self-referential twist, the T-1000 as a technology beyond computers is represented with the most advanced computer graphics available. The audience, which knew this very well from advance publicity, admired the T-1000's computer-based form even as it participated in constructing a postcomputational, postcyborg Other.) For most of the film, however, the T-1000 appears as a police officer, in uniform and driving a police vehicle. Its first act is to type John Connor's name into the police car's computer, which instantly provides a detailed dossier. As also happens in *T1*, the enemy impersonates the police, turning the information and resources of the state against the hero.

If *T1* is about the reconstruction of female identity in the shadow of nuclear war, *T2* is about the search for parents in a world where traditional motherhood has vanished and fathers are deadbeats. We first encounter the ten-year-old John at the home of his harried, slightly sleazy, very ordinary foster parents. He ignores his foster mother's nagging. When Todd, her husband, admonishes him to "pay attention to your mother," John retorts, "She's not my mother, Todd" and roars off on his motorbike.

In the next scene we rejoin Sarah Connor and learn where women who abandon womanhood to fight the mechanized, computerized, nuclear future end up: in insane asylums. John's "real" mother awaits the holocaust in a maximum-security mental hospital—doing pull-ups in her cell, sweating, buffed up into a female version of Schwarzenegger, her lip curled in derisive anger. When her psychiatrist plays back a videotape of Sarah recounting her recurring nightmare of nuclear war, we find ourselves—we who have been told, in the post-Cold War world, that *this* nightmare at least is over—uncertain what to make of her uncontained rage.

John soon confides to us his conflicted relationship with Sarah. He proudly announces that he learned how to hack into ATMs from his "real" mom. But he immediately tells his friend, "She's a complete psycho . . . a total loser." Later, we learn more of John's family history:

> We spent a lot of time in Nicaragua and places like that. . . . For a while there she was with this crazy ex-Green Beret guy. Running guns. Then there were some other guys. She'd shack up with anybody she could learn from so she could teach me to be this great military leader. Then she gets busted. It's like,

> sorry kid, your mom's a psycho. Didn't you know? It was like everything I'd been brought up to believe was all made of bullshit. I hated her for that.

Sarah, like revolutionary women throughout history, is portrayed as prostituting herself for the cause. The cause has turned out, as usual, to be a false promise, an illusory struggle, doomed to failure. John, child of the age of single working mothers, serial monogamy, and "family values," rejects both his ersatz nuclear family and his insane "real" mother.

But once the film reveals which Terminator is the villain and which the protector, John sets off with the T-800, pursued by the T-1000, to rescue Sarah.

When they arrive at the asylum, she has just been recaptured after a valiant but unsuccessful attempt to break out. The T-800 extends his hand to her, saying sternly, "Come with me if you want to live." The terrified Sarah of 1994 is confronted with a seeming reprise of her 1984 nightmare. She is faced with the wrenching task of transforming her image of the cyborg as monster into a new picture, the cyborg as protector. To do so, she must acknowledge his subjectivity as familiar. In the car after their escape, she turns to the T-800 and offers: "So what's *your* story?"

Now the film begins its central ideological task, the reconstruction of motherhood, fatherhood, the nuclear family, and the white male as world savior. John's role in the process of rehabilitating Sarah as traditional woman and mother will be to provoke her suppressed emotionality.

John is also given the task of reconstructing a father figure from the mechanical material of Schwarzenegger's T-800, which, like a robotic genie, must obey John's direct commands. John learns that the T-800 is not, after all, a rigidly programmed, single-minded monster, but a creative, learning subject who feels an analog of pain. "Can you learn stuff that you haven't been programmed with, so you can be, you know, more human, and not such a dork all the time?" "My CPU is a neural net processor—a learning computer," the T-800 replies. "The more contact I have with humans, the more I learn." John: "Cool!" The abstract computation of symbolic AI is replaced by the biofriendly neural network and its learning capacities, in keeping with the cyborg's new aspect as friend and protector.

As the film proceeds, John steadily trains this awkward, overserious cyborg in human speech, human relations—and human emotions. The T-800 asks why people cry. John explains that "it's when there's nothing wrong with you, but you hurt anyway." John teaches the T-800 not to kill people; he settles for grievously wounding them instead. We are meant to see the cyborg as a kind of National Rifle Association-issue protector who uses violence willingly but only when "necessary." Like the Persian Gulf battles, with their managed popular perceptions of war as a bloodless computer game, the new cyborg figure is pure New World Order.

At Sarah's survivalist hideout in the desert, they talk and play. Sarah, watching them, sees what is happening: "Watching John with the machine, it was suddenly so clear. The Terminator would always be there, and it would die to protect him. Of all the would-be fathers who came and went over the years, this thing—this machine—was the only one who measured up. In an insane world, it was the sanest choice."

The scene closes with John and the T-800 cavorting, cyborg father and human son, warmed by Sarah's bemused, nearly beaming gaze.

But the problem of nuclear war is not yet resolved. The T-800 explains how it will happen without a Cold War. The unsuspecting computer engineer Miles Dyson of Cyberdyne Systems "creates a revolutionary type of microprocessor" by reverse-engineering the broken computer chip found in the head of the original Terminator. "All Stealth bombers are upgraded with Cyberdyne computers, becoming fully unmanned. Afterwards they fly with a perfect operational record. The Skynet [computer] system goes on line on August 4th, 1997. Skynet begins to learn at a geometric rate. It becomes self-aware. . . . In a panic, they try to pull the plug. . . . Skynet launches its missiles against the targets in Russia." John wants to know, "Why attack Russia? Aren't they our friends now?" "Because Skynet knows that the Russian counterattack will eliminate its enemies over here."

At this point, Sarah drifts into sleep, into her nightmare of nuclear war. Watching a playground, she sees herself through the chain-link fence as a young mother playing with her toddler, a massive cityscape in the distance behind them. A nuclear blast destroys it all, everything "breaking apart like leaves." She wakes from the nightmare driven, and roars off alone to kill Dyson in order to prevent the war. A few minutes later, John realizes where she has gone, and pursues her with the T-800.

Now it is time for Sarah's final rehabilitation. She finds Dyson working at home with his wife and young son—and she attacks him with an automatic rifle. In a gut-wrenching scene, she strides into the family home, where the trembling Dyson lies bleeding in his wife's arms, to finish him off. "It's all your fault," she screams. But Sarah, a "woman" after all, cannot bring herself to fire the fatal bullet. In the face of this mirror image of her own recently reconstituted "family," she crumples to the floor, sobbing. At this juncture, John and the T-800 arrive. John takes over the scene.

This is the ultimate moment of redemption. Sarah sobs, "I almost . . . I almost . . . [killed him]." Sobbing even harder in John's young arms, she cries, "You came here to stop me. I love you, John. I always have." Her emotionality, her vulnerability, her need for a male savior restored to her, Sarah recovers her full womanhood.

In the aftermath, the film gently but firmly repudiates the radical feminism of Sarah's viewpoint. When Dyson wonders how he could have known where his research would lead, Sarah sneers, "Fucking *men* like you built the hydrogen bomb. Men like you thought it up. You think you're so creative. You don't know what it's like to really create something—to create a life, feel it growing inside you. All you know how to create is death . . . and destruction." As she builds to a crescendo of male-blaming rage, John interrupts her. "Mom! We need to be a little more constructive here." In the name of pragmatics, he effectively silences the feminist critique of both male science and the gendered institution of war. We are clearly meant, between this and other scenes of Sarah's intense brooding, to wonder whether she is not at least a little bit crazy after all.

After all this ideological activity, the film's final scenes are almost anticlimactic. Dyson, a good family man, proves more than willing to take personal responsibility for his actions. He goes with Sarah, John, and the T-800 to the Cyberdyne plant to destroy Cyberdyne's files and the original Terminator microchip. After one final chase, Sarah nearly succeeds in destroying the T-1000, as she had crushed the original Terminator in *T1*. But not quite. This time her own strength is not enough, and the damaged (male) T-800 reappears just in time to save her and John by blasting the T-1000 into a boiling vat of molten steel, reconstituting male protectorship.

In a maudlin final scene, the fantasy of the superhuman father comes to a bitter end. Although the old chip and the T-1000 are destroyed, the T-800 reminds them that "there is one more chip"—pointing to his own head—"and it must be destroyed also." John resists, crying. But the cyborg must be sacrificed—he cannot, finally, become the perfect father John and Sarah both desire. Mirroring the T-800's gesture, Sarah offers him her hand. As they clasp, she nods in respectful recognition. She then lowers him into the steel pit, looking sadly after him as he descends, while John watches and cries. In the Terminator's final moment we see, for the last time, the Terminator's-eye screen display as it crackles with static, fades to a single dot and then goes out.

We are meant to mourn his death.

Conclusion

What messages can we decode in these snapshots of the politics of cyborg subjectivity?

T2's rehabilitation of the cyborg reflects a theme common in science-fiction film from the mid-1980s on. In *2010* (1984), the sequel to *2001: A Space Odyssey* (1968), Stanley Kubrick's killer computer HAL is revealed to be the victim of mental illness induced by self-contradictory programming. In arresting moments of intimate communion between human and machine, the computer scientist Dr. Chandra heals HAL, who then sacrifices himself to save the human spaceship crew. Ridley Scott's 1979 *Alien* featured treachery by an android masquerading as a scientist. But in its 1986 sequel, *Aliens*, another android sacrifices himself to save the embattled heroine from the insectoid aliens. The enormously popular TV series *Star Trek: The Next Generation* foregrounds the android Commander Data, with his bland, androgynous appeal and his puzzled attempts to comprehend human intuition and emotion. In all of these cases, respect for the cyborg's subjectivity goes hand in hand—as it were—with his (in these cases) relegation to the role of servant to human needs.

The nature of the subjectivity of cyborg figures in these films suggests that intelligent machines are being integrated into contemporary culture under the all-inclusive rubric of multiculturalism. Rather than a threat, their minds represent just one more curiosity for the anthropologically inclined, one more reminder of the diversity of sentient beings. At the same time, their role as servants points to a renewed sense of human control over autonomous technology. This sense is perhaps associated with the perceived demise of threats of nuclear Armageddon as well as with the widespread acceptance of computer-based automation and computerized communication as inevitable concomitants of modern economies.

T2 seems to suggest that we are waking up from a nightmare run by cyborgs to a world we can control. Computers that learn and create, it hints, can help us rebuild a shattered social world in the image of the past. The figures of the violent, unfeeling, but devoted father-protector and the emotional, unstable, but loving mother return to form the basis of the post-Cold War social order. The cyborg-machine returns to its place as technological servant.

But some transformations of identity remain. Sarah, even with her emotionality and male dependence, is still far tougher than most of the men she encounters. Only the scientist and father Dyson, in taking personal responsibility for the holocaust, matches the Terminator's manhood. The Terminator, like every would-be father before him, finally leaves the scene. He turns out to be merely a fantasy, since the future that produces him is erased along with the Cold War. Despite its ideological retreat, *T2* remains profoundly distrustful of white men. It leaves us, in the end, without a great white father. (Dyson is black.)

Can the cyborg figure, as Haraway understands it, still serve as a potent resource for the reconstruction of gender and other political-cultural identities along the technology/biology divide? How will it evolve in a world where commercial goals replace military support as the fundamental drivers of advanced computer and information technology? It is too early for definitive answers to these questions. But the integration of cyborg figures in fiction points to one of the fundamental problems cyborg politics may now encounter.

A simplistic multiculturalism flattens all difference into two categories. *Exotic* differences function as the "interesting" resources for a Ripley's Believe It or Not pseudoanthropology. *Mundane* differences form the basis for an easy relativism that avoids hard ethical and epistemological questions by rotating all hierarchies—whether of value, of knowledge, or of power—sideways by ninety degrees. Both of these ways of conceiving (and perceiving) cultural differences ignore the really pressing problems of a multicultural world: finding new sources of integrity and authenticity in a world without genuine traditions, preserving cultural diversity under the gigantic pressures for integration induced by modern economies and modern media, and locating a basis for politically expensive judgments of value in the face of the political cheapness of moral relativism.

Haraway's understanding of cyborg identity does not succumb to these failings; it is antifoundationalist but situated, never fully relativistic.[19] Yet *Terminator 2* suggests the possibility that in the New World Order cyborg politics may be co-opted by a simplistic multiculturalism, transformed into exoticism and, increasingly, into merely mundane differences between sentient creatures whose technology matters no more than their biology. If the cyborg figure really does offer a challenge to essentialism, it also challenges us to find new anchor points and new coherences—ones that remain cognizant of that figure's origins as the automated technological soldier of the Cold War—in order to go beyond it.

Nevertheless, here in the New World Order we—for in the end we are cyborgs ourselves—may find at least a glimmer of hope in the demise of Cold War politics, with its grand ideologies and subordination of culture to war. Rolling along down a night highway, Sarah reflects on what lies ahead: "The unknown future rolls toward us, and for the first time I face it with hope. Because if a machine, a Terminator, can learn the value of human life, maybe we can too."

Notes

1. Donna J. Haraway, "A Manifesto for Cyborgs: Science, Technology, and Socialist Feminism in the 1980s," *Socialist Review* 15, no. 2 (1985): 65-107.
2. Paul N. Edwards, "Are 'Intelligent Weapons' Feasible? DARPA's Billion-Dollar Bet," *Nation* 240, no. 4 (1985): 110–12; Paul N. Edwards, "A History of Computers in Weapons Systems," in *Computers in Battle*, ed. David Bellin and Gary Chapman (New York: Harcourt, 1987), 45–60; Paul N. Edwards, "The Army and the Microworld: Computers and the Militarized Politics of Gender," *Signs* 16, no. 1 (1990): 102–27; Paul N. Edwards, *The Closed World: Computers and the Politics of Discourse in Cold War America* (Cambridge: MIT Press, 1996); Chris Hables Gray, "Computers as Weapons and Metaphors: The U.S. Military 1940-1990 and Postmodern War" (Ph.D. thesis, University of California, Santa Cruz, 1991). Much of this essay is adapted from Edwards, *The Closed World*.
3. Fred Halliday, *The Making of the Second Cold War* (London: Verso, 1986).
4. David Bellin and Gary Chapman, eds., *Computers in Battle: Will They Work?* (New York: Harcourt Brace, 1987); Barry Bruce-Briggs, *The Shield of Faith: Strategic Defense from Zeppelins to Star Wars* (New York: Touchstone, 1988); James Chace and Caleb Carr, *America Invulnerable: The Quest for Absolute Security from 1812 to Star Wars* (New York: Summit Books, 1988).
5. Alan Borning, "Computer System Reliability and Nuclear War," *Communications of the ACM* 30, no. 2 (1987): 112–31; Paul Bracken, *The Command and Control of Nuclear Forces* (New Haven: Yale University Press, 1984).
6. Edwards, *The Closed World*.
7. Ridley Scott also directed the science-fiction cult films *Alien* (1981) and *Blade Runner* (1984).
8. Edward Feigenbaum and Pamela McCorduck, *The Fifth Generation: Japan's Computer Challenge to the World* (Reading, Mass.: Addison-Wesley, 1983); Edward Feigenbaum, Pamela McCorduck, and H. Penny Nii, *The Rise of the Expert Company: How Visionary Companies Are Using Artificial Intelligence to Achieve Higher Productivity and Profits* (New York: Times Books, 1988).
9. Hubert Dreyfus and Stuart Dreyfus, *Mind over Machine* (New York: Free Press, 1986); Harry Collins, *Artificial Experts: Social Knowledge and Intelligent Machines* (Cambridge: MIT Press, 1990).
10. Douglas Lenat, Ramanathan Guha, Karen Pittman, Dexter Pratt, and Mary Shepherd, "Cyc: Toward Programs with Common Sense," *Communications of the ACM* 33, no. 8 (1990): 31–49.
11. Helen Rogan, *Mixed Company* (Boston: Beacon Press, 1981); Judith Stiehm, "The Protected, the Protector, the Defender," in *Women and Men's Wars*, ed. Judith Stiehm (New York: Pergamon, 1983), 367–76.
12. On computers as marginal objects, compare Sherry Turkle, *The Second Self: Computers and the Human Spirit* (New York: Simon and Schuster, 1984).
13. This point emerged in conversation with Donna Haraway.
14. Theodore A. Postol, "Lessons of the Gulf War Experience with Patriot," *International Security* 16, no. 3 (1991–92): 119–71.
15. J. L. McClelland, D. E. Rumelhart, and G. E. Hinton, "The Appeal of Parallel Distributed Processing," in *Parallel Distributed Processing*, ed. J. L. McClelland and D. E. Rumelhart (Cambridge: MIT Press, 1986), vol. 1, 3–44; Paul Churchland, *A Neurocomputational Perspective* (Cambridge: MIT Press, 1989); Andy Clark, *Associative Engines: Connectionism, Concepts, and Representational Change* (Cambridge: MIT Press, 1993); Philip E. Agre and David Chapman, "What Are Plans For?" MIT Artificial In-

telligence Laboratory, *AI Memo 1050*, September 1988; Philip E. Agre, *The Dynamic Structure of Everyday Life* (Cambridge: Cambridge University Press, 1996); Rodney A. Brooks, "Intelligence without Representation," *Artificial Intelligence* 47 (1991): 139–59.

16. Statistics compiled in Win Treese, "The Internet Index," http://www.openmarket.com/diversions/internet-index.
17. William Gibson, *Neuromancer* (New York: Ace Books, 1984).
18. Michael L. Dertouzos, Vinton G. Cerf, Lawrence G. Tesler, Mark Weiser, Nicholas P. Negroponte, Lee Sproull, Sara Kiesler, Thomas W. Malone, John F. Rockart, Alan C. Kay, Al Gore, Anne W. Branscomb, and Mitchell Kapor, "Communications, Computers, and Networks: How to Work, Play, and Thrive in Cyberspace," special issue of *Scientific American* 265, no. 3 (1991).
19. Donna Haraway, "Situated Knowledges: The Science Question in Feminism and the Privilege of Partial Perspective," *Feminist Studies* 14, no. 3 (1988): 575–99; Donna J. Haraway, "The Promises of Monsters," in *Cultural Studies*, ed. Lawrence Grossberg, Cary Nelson, and Paula A. Treichler (New York: Routledge, 1992), 295–337.

Bodies of Knowledge:
Biology and the Intercultural University

Scott F. Gilbert

We are not seeing the end of the body, but rather the end of one kind of body and the beginning of another kind of body.
▸ *Emily Martin ("The End of the Body?" 1992)*

Our Selves/Our Bodies

In the academic debate about multiculturalism, the natural sciences are usually left out.[1] This is a gross error. Given the roles that science plays in our cultural life, its importance as a culturally informed artifact, and its role as an export product, our discussions of multiculturalism are superficial if we do not attend to the conditions that permit and direct scientific inquiry. This becomes all the more obvious in the case of biology, which has traditionally been central to our definitions of race, class, and gender and which structures our discussions of health and ecology. In this essay, I hope to present a framework for such a discussion of biology's role in multicultural studies, grounding it in the larger context of the body politic. The body politic paradigm allows us to bridge the space between our discussions of physical bodies, bodies of knowledge, and social bodies, and it will enable us to assimilate (digest?) new elements into the curriculum that "embodies" our academic ideals.

In illo tempore, once upon a time, there was a body. This body was the neural body. It thought, and therefore it was. Indeed, this body knew who it was and what it wanted. The brain was the command center of this embodiment, and the nervous system empowered the rest of the body to do its bidding. The neural body was structured on strictly Cartesian coordinates, and the Y axis was the Great Chain of Being. At the apex resided the brain with its God-given rational soul and self-knowledge. At the base lay the genitals. The heart of man was torn between these two poles: reason and emotion; intellect and passion; the *mens* of God and the *mentula* of Satan; the sphere of divine spirit and the corruption of base matter. Through nerves and hormones originating at the brain stem, the head dictated the production regimen of the body. This body is the Fordist body whose death has been proclaimed by Jonathan Parry and Emily Martin.[2] The neural axis was also seen as the axis of evolution, and we cerebro-manuals were at its apex. Like the brain atop the spinal cord, we felt ourselves positioned at the control center of nature.

Sociological discussions of the body metaphor assume that science has a single and uniform concept of the body, and it is usually the standard neural body that is the basis for the body politic analogy. However, after World War II, the authority of the neural body has become shared with two other views of the body, two other claimants to "selfhood." In addition to the neural body, there now exist the immune and genetic bodies. Each of these bodies privileges a different notion of identity ("self") that corresponds to a different type of body politic.[3] When one makes the "body politic" or "body of knowledge" metaphor, one is extrapolating a particular type of body into the social or academic sphere. The neural body privileges a polity defined by laws, mores, and culture and corresponds to the traditional aperspectival view of science. The immune body privileges a polity defined by defensible borders and corresponds to a view of science wherein science is a product of Western culture that preserves and defends the interests of the dominant groups of that culture. The genetic body privileges ethnicity and race as the bases for polities and corresponds to a view of science that sees science as the mainspring and major determinant of culture. In addition, there is a fourth "phenotypic" body that attempts to mediate and integrate the other three. Whereas the first three form our major images of the body, the fourth stands opposed to the others as a counterimage.

The body-derived metaphors are among the most central to our perceptions of reality,[4] and each of these views of the body has important consequences when extrapolated into political and social spheres. I hope to demonstrate that our perceptions of social problems and our views of science depend to a great extent on which bodily concept is being employed. Moreover, the ascendancy of the genetic body in recent years has potentially far-reaching effects on our science curriculum and our discussions of social change.

The Neural Body

When the notion of the neural body eventually died, its flesh deconstructed about it. The body resurrected from its ashes is no longer the neural body. I will argue that there are now four selves in the present image of the body, each with its own nature and center. There is, of course, the reconstructed neural body. Although it retains many of its nineteenth-century characteristics, it is now a much more mature body that privileges the *interactions* between the brain and the other organs. The monarchy has moved from being absolute to being constitutional. The monarch rules by the consent of the governed and is not uninfluenced by the heart, guts, gonads, and kidneys. The contemporary neural body is an integrated system, but the brain—with its powers of thought, memory, and volition—is still seen as the center of identity. If one were to transplant a brain into a new body, the self-identity would stay with the brain.

The Immune Body

However, since World War II, three other bodies are now distinguishable. These bodies had always been part of our consciousness, but they were considered subsidiary and servant to the neural body. Our second body is the immune body. This is a body that separates us from the other bodies that inhabit the globe and that prohibits our fusing with other entities. The immune body is that which determines our Hobbesian selfness and is in potential conflict with every other body. I am not you, and were your body tissues to be grafted to mine, the graft would not take.[5] Indeed, immunology has been described as "the science of Self-Nonself discrimination."[6] The boundary for the immune body is not the brain, but the skin (which rejects transplants). There is no "center" analogous to the brain in the immune body. Rather, the cells of the immune system pervade the entire body.[7]

This immune body had not been noticed before the 1940s because there did not seem to be any alternatives to the type of description used for the body. Whereas we see the "dumb" animals all around us, and therefore can recognize our neural bodies as being different from theirs, we did not have any other model for the immune body. The immune body was not an independent entity, but merely a way of preserving the integrity of the neural body. The neural body was pure but susceptible to outside noxious infections. This was reflected in our concern over the diseases that threatened our body politic. They were infectious diseases: we "quarantined" Cuba and fought "Communist infiltration." During the McCarthy era, people could become contaminated by ideas.

Only in the 1960s, when views of what was internal and external to the self underwent a major revolution, and when antibiotics and vaccines had alleviated most social fears about infection in the Northern Hemisphere, did the prevalent notion of the immune self change.[8] The disease of the body politic went from infection to cancer. In this latter trope, the immune self is very different. Unlike Susan Sontag,[9] I see a radical break between the metaphors of infection and cancer. The infection model assumes that the body is initially pure, and is susceptible to noxious, external agents. The cancer model made no such assumption. In fact, it posited the opposite. The danger to the body is that a normal part of the body has escaped its normal regulatory network and has become autonomous. The danger was not from without; it was from within. Cancer was a disease that reflected lack of control. Something that was normal for the body had escaped regulation and was expanding unchecked—the military-industrial complex, media, permissiveness, youth. As Walt Kelly remarked, "We have met the enemy, and he is us."

But cancer did not remain the predominant disease mode within the body politic metaphor. In the late 1970s, as the population of the Northern world was growing older, the model of cancer gave way to a model of decrepitude and senescence. The infrastructure is crumbling. Expect less. Like Greece, Rome, and Great Britain before it, America had had its day in the sun and was now in decline. We were thinking recessionally, and we became fascinated with the passage of the British Raj (and were treated to movies such as *Ghandi*, *The Jewel in the Crown*, and *A Passage to India*). Newspapers and magazines were full of stories of bridge collapses, tunnels needing repair, and the withering away of our railroad, highway, and airline systems. Although portions of the cancer (and to some degree the infection) trope still continued to inform our notion of the "disease of the body politic," decrepitude and senescence came to play a larger and more substantial role.

This concern with the decline of the body politic was mirrored in our concern over the physical body; for our physical body was also in decline, and the problem was not the fast-growing cancers, but the slow but steady collapse of our own infrastructure. In the early 1980s, there were twice as many Americans over the age of sixty-five as there were as late as 1950, and those baby boomers who would never trust anyone over thirty were turning gray themselves. Jane Fonda switched her activism from international politics to the means of preventing bodily decay. Even former campus radicals became concerned with having a fit body. The specific debilitating disease we chose to mirror our anxieties was a particularly frightening one. From the late 1970s to the early 1980s, there was a large increase of interest in Alzheimer's disease. Alzheimer's disease reflected very intense anxieties. Not only were we getting older physically, but our minds were going as well. We had no history, no memory of the past.

This Alzheimer's model did not last long, because the 1980s saw a new disease of the immune body—AIDS.[10] AIDS combines several tropes of the disease of the body politic. First, it is seen as a disease that features decrepitude. Like Alzheimer's disease, AIDS is a disease of wasting away. Look at the pictures of Rock Hudson before and during AIDS. Recall the photographs of all those young people in wheelchairs. The decrepitude metaphor is clearly a critical part of the AIDS trope. Second, AIDS is a disease characterized by lack of control over one's bodily identity. People with AIDS find their bodies unable to separate from the world around them. Organisms that had been external become internal. The peripheral is allowed to become central. Third, the AIDS trope contains an infectious component, the nidus of which is in the Third World enclaves (inner cities and, originally, Haiti) within the Northern body politic. Just as an infection is seen as becoming part of the body, so the metaphor can be transferred to those people constituting these enclaves.[11] Thus, the AIDS model resonates between *(a)* the cancer metaphor, because it stresses what is internal to the body politic, *(b)* the Alzheimer's model, because it stresses decrepitude within the body politic, and *(c)* the infection model, because it can identify elements of the "neural" body politic that were originally "foreign" to it and that are threatening to it. The neural body is not the only "body" we have. The immune body is obviously important when it comes to defining our political relationships. But these two are not our only bodies.

The Genetic Body

Our third body/self is the genetic body. We are what (who?) our genes tell us we are. It is our repertoire of phenotypes. The body is seen as but an epiphenomenon of its genes. The center of this body's identity is not within the brain or the lymphocytes. Rather, it lies within the nucleus of every somatic cell of the body. This is the identity of the body sought by the Human Genome Project. It is the identity of the body celebrated by the antichoice brigades who tell us that science has fixed the moment of individuality at the moment of conception. Whereas the neural and immune bodies can learn and change, the genetic body cannot. This gives it a more primal immediacy than the neural or immune bodies. The discovery of new human genes makes the front page of newspapers because of the claim that the genes determine who we are.

The radial axis of the genetic body is, as in the neural body, the Great Chain. The chromosomes of the nucleus are perceived as being the blueprints—the idea—upon which the body is constructed. It is the "executive suite of the cell,"[12] while the cytoplasm is but the factory floor. The nucleus is unchanging, while the cytoplasm is always bringing in new matter and getting rid of the old. Genes are the unmoved movers of the bodily cosmos. Although the bodily substance changes constantly through metabolism, the identity of the body stays the same thanks to its unchanging nuclei. Like the neural body and the immune body, the genetic body concerns information, not substance. Dorothy Nelkin and M. Susan Lindee have shown the similarities of the genetic self to the medieval soul.[13] Not only does our genome determine our physical and psychological characters, but it can survive the death of the body and be the basis for bodily resurrection. This is what *Jurassic Park* was all about.

Needless to say, these concepts of the body are not coincident. Indeed, they provide three distinctly different representations of the body. This is also seen in the late-twentieth-century concepts of the body politic. Is the body politic the genetic, immune, or neural body? That is: Is a body politic, such as a country, defined by the ethnicity of its people (genetic), its ability to defend itself against neighboring countries (immune), or its culture and mores (neural)? Similarly, in the debate over abortion, is the self to be defined genetically (when the nuclei fuse at conception), neurally (when the EEG pattern begins around the seventh month), or immunologically (when the separation between mother and baby occurs at birth)? Interestingly, it is only very rarely (if at all) that these three concepts have the same borders, and most of the wars presently being waged (for instance, in Russia, the United Kingdom, the Near East, and Africa) are due to the nonalignment of these boundaries and the belief that all three should be in register.

The Phenotypic Body

The fourth self stands in contrast to the other three. This is the phenotypic body. It is all these three bodies plus the flesh, bones, organs, and tissues. It is the lover's body of flesh and substance. It is the cripple's body; it is the body of all those undergoing marked physical changes. This is the body created by those sciences that celebrate reciprocal interactions and emergent properties of systems—embryology and ecology.

There has been a long history of embryologists (notably Oscar Hertwig, N. J. Berrill, P. Weiss, and C. H. Waddington; presently B. C. Goodwin, H. F. Nijhout, and myself, among others) who have argued against the linear, univectoral, notion that genes determine the character of either cells or selves. The fate of a cell is determined in the body by its interactions with other cells, and we are walking on cells that could have been used for thinking had they been in another part of the embryo. Rather than genetic determinism (or neural determinism), we have proposed that nothing is determined except by interactions. Hertwig, Alfred Tauber, and Scott Gilbert and Steven Borish explicitly use this "epigenetic" approach to model both the body (which forms embryologically by cell-cell interactions) and the self (which forms epigenetically through interactions with others), in addition to the autonomous directions given by the nuclear genes.[14]

However, contemporary biology is largely controlled by the genetic model of life, and even neurobiology and immunology have been redefined as gene-based sciences. This transition from being phenotypic sciences to being genotypic sciences was made even earlier for physiology, embryology, and evolutionary biology.[15] Those that resist this "molecularization" (notably ecology, paleontology, and some aspects of embryology and taxonomy) are marginalized. Moreover, we still have taboos about discussing the physical body and prefer to frame our discourse around the other three selves. Consider, for example, contemporary discussions on homosexuality. One would expect that such discourse would be primarily informed by the sexual body. Instead, one finds arguments concerning "gay genes," "gay brains," and, of course, the "gay" immune system. The same can probably be said for most academic discussions of women (although I suspect that in both cases this is changing). Nevertheless, this interactive, phenotypic, continually self-creating body is one where power relationships have diminished and where an organicist wholism prevails: the body as an integrated network of gene and flesh, brain and gonad, inside and outside.

At this moment, there are fascinating areas of genetics, immunology, and neurobiology that are returning to epigenetic interactive models.[16] This is especially true of immunology, where the immune network theories have revised the internal-external system that had been the hallmark of immunology and that remains its main legacy to political literature. As it stands now, though, this phenotypic body is not usually a major component of our corporate self; rather, it is perceived as an epiphenomenon of the other bodies.

Four Bodily Representations of Science

Presently, there exists a tension between natural science and intercultural learning that is reflected in Sandra Harding's dictum, "Science is politics by other means, and it also generates reliable information about the empirical world."[17] How can sciences become part of a multicultural curriculum? Here, the body politic metaphor can be extremely helpful, for the sciences are said to be (among other things) "bodies" of knowledge. But what type of body? I would depict science as oscillating/resonating among the four types of bodies discussed earlier, and each of these bodies privileges a different relationship between science and the larger culture. Like the physical and social bodies, science is all four of these bodies simultaneously, and any view of science that does not include all of them is incomplete. By seeing science as oscillating/resonating among these four bodily representations, we can see how natural science can become an integral feature in a multicultural curriculum.

Science as a "Neural" Body of Knowledge

The neural view of science is that science finds acultural truths. In this body of knowledge, science is disembodied thought, and it is seen to provide knowledge that is not culturally situated. This aculturalism is the flip side of multiculturalism. The truths of science are independent of their material circumstances. Laws of science should be true anywhere on Earth, indeed, anywhere in the universe. The laws should be ahistorical, acultural, and apersonal. The simplicity and starkness of the laboratory report originated in the attempt to rid science of cultural specificity, thus allowing devout Protestants such as Newton, Kepler, and Boyle to use the insights and data of Catholics such as Copernicus, Descartes, and Galileo.[18] Given that the conclusions are acultural, science welcomes anyone into it. As Galileo exclaimed, "Anyone can see through my telescope." The authors of scientific papers are supposed to be culturally interchangeable. It does not matter whether R. E. Lee is a Virginia gentleman or a Chinese woman, since the methods used should eliminate any cultural biases. The list of Nobel Prize winners in science is a wonderful mix of ethnic names and reflects the inclusiveness of science. This has been the traditional view of science; and "bad" science is said to result from the imposition of cultural norms on objective investigations.

Science as an "Immune" Body of Knowledge

In the immunological view of science, science is used for the construction, maintenance, and defense of the body politic. Here, science is a prime example of Western culture; despite its belief that it is universal, it is actually extremely parochial.[19] First, the notion of an acultural, ahistorical, apersonal law is a distinctly Western idea, derived from the scriptural traditions of a transcendent deity whose laws run the universe even in the absence of human beings. Such a universal and apersonal law is a foreign concept to most traditional cultures. The origin and early development of modern science is closely intertwined with the emergence of modern society in the West, and, as Sal Restivo has summarized, "The scientific revolution was one of an interrelated set of parallel organizational responses within the major spheres of Western Europe from the fifteenth century onward (including Protestantism in the religious sphere and modern capitalism in the economic sphere) to an underlying set of ecological, demographic, and political economic conditions."[20] Thus, contemporary science is a Western product grounded in Western sociopolitical views of nature. Second, science fails to meet its ideal of aculturality. Obvious cases have been social Darwinism, scientific racism, scientific antifeminism, and Aryan physics. Science often defines race and gender in light of the opinions of scientists who, until very recently, have been overwhelmingly male, economically privileged, and white. At the beginning of this century, American and European biologists declared that there was a graduated scale of humanity, and that certain anatomical features showed that black males were on the same level as white females and white children—that is, lower than that of white men. Moreover, some of these texts suggested that we treat such people as we would treat animals.[21] In 1972, the major textbook in mammalian developmental biology contained the following passage: "In all systems that we have considered, maleness means mastery. The Y chromosome over the X, the medulla over the cortex, androgen over estrogen. So physiologically speaking, there is no justification for believing in the equality of the sexes."[22]

Less obvious politicizations of biology can be seen in the sexist and classist description of cells. There is a history of depicting the nucleus (and sperm) as active, rational, and executive, while the cytoplasm (and egg) are depicted as passive, changeable, and proletarian.[23] Western cultural norms are also seen reflected in the notion of self used in evolutionary theory.[24] Thus, science reflects the dominant views of those who created it. It has been used to sanction and to protect the social order of the Western world from which it comes.

Science as a "Genetic" Body of Knowledge

In the genetic body of science, science is seen as part of culture that has achieved independence and is now the controlling element of the culture. In recent years, several scholars have reiterated C. P. Snow's view that science has become a culture unto itself (albeit with definite links to its parent culture), and that many "educated" people are totally ignorant of this culture. Here, the proper study of humankind is science, and science is "the center of culture."[25] According to Alvin Toffler and Heidi Toffler, science (and the technology derived from it) is the defining aspect of our civilization.[26] Indeed, science is what we humans are all about. If one were to ask what is the critical trait that distinguishes humans from animals, the answer is that humans do science. One author puts the proposition as follows: "If humans exist on earth for a purpose, it is likely to be for scientific research. It is the one urge that is exclusively human and distinctive of the race."[27]

According to this view, science should be taught as a culture just like any other. Again, Sandra Harding: "We live in a scientific culture; to be scientifically illiterate is simply to be illiterate—a condition of far too many women and men."[28] Science is a creative human endeavor, and some of the conclusions of science (such as the periodic table, the laws of motion, the laws of relativity, the laws of gravity and electromagnetism, the laws of thermodynamics, the laws of heredity, the cell theory, the gene theory, and the theory of evolution) are among the most creative acts of the human species. Everyone should consider them as part of our heritage.

However, a corollary to this is that science has become the most important part and the controlling entity of our culture. Within the past two decades, we have gotten our computers, fax machines, E-mail, microwave ovens, walkmen, discmen, boom boxes, CD players, credit bureau databases, airline booking systems, in vitro fertilization clinics, and video recorders. Meanwhile, jets, antihistamines, tranquilizers, air conditioning, car stereos, antibiotics, oral contraceptives, and hydrogen bombs are all things we (or at least our children) take for granted that did not exist prior to World War II. What else has increased (or improved) so greatly over those years: our art, music, philosophy, or literature? The thing that distinguishes our culture from all others is its science. Is it any wonder that it is so envied and feared?

Science as a "Phenotypic" Body of Knowledge

The phenotypic view of science is isomorphic to the phenotypic conception of the self. It values matter and practical utility as more important than abstract theories. Such a science would see biology, chemistry, and physics as steps toward achieving the ends of medicine and engineering. As revolutionary as this may sound, historians of science have long understood that science had two fonts: Descartes's idea of the comprehensive understanding of nature, and Bacon's idea of serving humanity through useful scientific knowledge. We would also do well to remember that biology justified its way into the liberal arts curriculum as a prerequisite for medical students.[29]

Descartes's ideal, of course, was mathematics, and going against this ideal means a major inversion in our understanding of science. The natural sciences, like most of the academy, are organized along the Great Chain of Being, the theoretical end being considered more important than the practical end. This appears to be true whether one is in the humanities, the social sciences, or the natural sciences. However, it is probably most pronounced in the natural sciences.

A few years ago, a student confided to me that she had trouble convincing herself to major in biology. Biology, she declared, was "so far down on the hierarchy" that she didn't think the best minds would work there. When asked "What hierarchy?" her answer was: "You know. There's math on the top, then physics, then chemistry, then, down at the bottom, biology." Of course. *That* hierarchy. We have all heard of that hierarchy, and many of us have accepted it and unthinkingly internalized it. What is it a hierarchy of? Very simply, it is the progression of matter into rationality. Whereas the Great Chain originally represented the extension of rationality over matter throughout the Universe, it now extends merely across the University. Biology deals with dirty matter: blood, guts, menstrual fluid, semen, urine, leaf mold, frogs, jellyfish, lions, tigers, and bears. Chemistry deals with purified and quantified matter: 20 mg/ml NaCl, 4 mM ATP. Physics deals with idealized matter (when it deals with matter at all): ideal gases, electron probability clouds, frictionless surfaces. (If physics deals too much with material, it falls down a branch of the Chain to become engineering.) Finally, mathematics claims to have escaped matter altogether! "In the pure mathematics we contemplate absolute truths which existed in the divine mind before the morning stars sang together, and which will continue to exist when the last of their radiant host shall have fallen from the heaven."[30]

In a similar vein, Oswald Spengler celebrated "the liberation of geometry from the visual, and of algebra from the notion of magnitude."[31] The levels of mathematics, it would seem, have replaced the orders of angels in our modern hierarchy. There are, as expected in such a Chain, bridge disciplines. As our students will tell us, biology spans from lowly ecology to molecular biology, and molecular biology at the upper end of the biology department touches biochemistry at the bottom of the chemistry department. Similarly, physical chemistry is seen as the highest chemistry, and theoretical physics the highest physics. Thus, as one rises up the Chain, matter becomes more and more abstracted until even the idea of matter is no longer present. So the organization and structure of science in the late twentieth century remains that of the Great Chain of Being, the traditional neural body politic.

As mentioned earlier, the three dominant views of the body all privilege information over substance, and only the "phenotypic" self privileges matter. Similarly, this alternative view of science exalts the more material sciences such as biology and the more practical parts of chemistry and physics.[32]

Science and the Multicultural Curriculum

Teaching Responsibility through the Neural, Immune, and Phenotypic Bodies of Science

There are numerous reasons for the inclusion of science in the multicultural curriculum. The first reason is obvious: the scientists being trained today are going to have powers far beyond anything in human history. They have to be taught responsibility. Here, the knowledge that science has an "immunological" body is important. Each science has its own history and its own problems: biologists have been responsible for advancing gender and racial hierarchies; chemists have a history of being unconcerned with the environmental effects of their molecules; physicists have to be wary about militarism. Scientists who know the history of their discipline—who know how their science supported racist, imperialist, and sexist agendas, who know how Nazi biopolicy was constructed, who know how eugenics was formulated—are less likely to repeat these errors. It is necessary to bring these issues to the science majors as part of their courses in the science departments. Lectures, laboratory readings, and discussion groups can all be utilized to this end. It is more important for everyone taking genetics to know how it was used in the 1920s than it is for a small group of students to learn about those episodes in their history of science courses.

Second, the "neural" body of science can be improved by the multicultural curriculum. Multicultural awareness can make better science. In the "neural" model, bad science is science that is informed by particular social norms and customs. The multicultural curriculum can locate these peculiarities. If science students realize that their own science has certain social assumptions, then it is not a large step to their realizing that by questioning some of the social assumptions they are questioning their science. However, the ability to question one's own social assumptions often means learning that one *has* assumptions, and that is done by knowing another culture. Knowledge of other cultures allows us to reflect on assumptions that we do not usually question. For instance, many gendered assumptions in biology were unquestioned until feminist critiques of biology became available.[33] Similarly, when viewed from Russian or Asian perspectives, several theories in evolutionary biology appear to be grounded in the particularly Anglo-Saxon views of selfhood and society.[34] Knowledge of these alternative concepts of self and society enable us to see the cultural specificities and limitations of our own theories. (Indeed, this present discussion of self and body is limited to the "Western" academy.) In these ways, a multicultural approach can train better-equipped scientists and create better science.

Third, science must be seen as part of our commerce with other cultures. Here, the "phenotypic" model of science is critical. Other cultures often had sciences, and the histories of these interactions are fascinating in their own right. But, more than history, our science can have serious effects on the other cultures of the world. In one course on biotechnology, the final exam consists of writing a grant application to apply modern biotechnologies to a problem in the "Third World." The students, of course, have to research what problems are important in such places. Then they have to find out how modern techniques could be implemented, if at all. Are they going to be acceptable to the people whom they are expected to help? Why should a particular problem be interesting to a particular culture? Why should certain problems be more important than others? The final examination becomes more than just an exercise in repeating knowledge. It can become a mode of discovery. Such courses show students that science is not done in a cultural vacuum and that science has its limits.

The Replacement of "Western Civ" by "Intro. Bio": The Genetic Body of Science

The genetic view of science states that science is the motor of our society. And this has far-reaching consequences that are already being realized. So far, our discussion of biology has limited it to a relatively small part of the curriculum. That has been its traditional place. But look again at your own school. Look at library budgets, look at enrollment figures, look at course offerings, look at the number of departments at universities. Biology is replacing Western Civilization. More specifically, "Intro. Bio" is replacing "Western Civ" as the course students must have to be informed members of the polity. Even if this statement is true merely at the curriculum level, I do not know if I am comfortable with this idea. However, I think it is coming upon us. The "Western Civ" course died in the 1970s,[35] and the period following its death has been characterized by confusion, especially in the humanities departments, and by the attempt to find a substitute and successor. None has yet been found. Intro. Bio is getting prepared to fill that niche.

The "Western Civ" course was one of the most consciously and politically constructed programs in all academia. Indeed, nothing could be further from the truth than the notion that the standard liberal arts curriculum is apolitical and that we are presently politicizing it. Rather, the "Western Civ" curriculum was founded on the basis of the War Issues Course initiated largely through the insights of a remarkable group of peace-loving professors who worked for the Department of War during 1918–20. Under the directorship of Dr. Frank Aydelotte of the Massachusetts Institute of Technology, this curriculum aimed "to educate recently conscripted American soldiers about to fight in France . . . to introduce them to the European heritage in whose defense they were soon to risk their lives. A new tie to Europe was instituted in relation to a national imperative." Historian Gilbert Allardyce continues, "More precisely, it was the child of a strange marriage between war propaganda and the liberal arts."[36]

This course (originally called the War Aims Course) was first given to the technical divisions of the army, and its model was a course that E. D.

MacDonald gave at the Wentworth Institute. (MacDonald became the assistant director of the program.) It consisted of a series of lectures on "the geographic, racial, economic, and political causes of the War; the responsibility of each nation involved; their ideals and aims, and the duty of the American soldier." As the War Studies Course, it was taught to every student enrolled in the Student Army Training Program, and it included "the historical background to the War, study of European governments, their ideals as set forth in their literatures and philosophies, and their aims." It was a mix of political science, literature, philosophy, and sociology.[37] After the Great War, Aydelotte (now president of Swarthmore College) strove to continue these courses in those schools that had taught them during wartime. He arranged with the World Peace Foundation and the National Board for Historical Service to provide funds for the textbooks and maps. Nearly every college and university responded positively. Aydelotte wrote that "I think the outstanding result of the War Issues course is the so-called 'Contemporary Civilization' course at Columbia, which is required of all Freshmen."[38] Other institutions were to follow Columbia's lead in creating such a course.

The War Studies Course had a heavy Anglo-Saxon bias to it, as might be expected of a wartime course celebrating the alliance of Britain and the United States and the rescue of European democracy by its American descendants. Aydelotte was an Anglophile in many of his academic and diplomatic pursuits. Indeed, the whole idea for the War Issues Course began when Aydellote "went over to Wentworth Institute, inspected their work, and lectured to a group of the men on our relationships with England and Anglo-Saxon ideals of democracy."[39] This love of British culture was not lost on Aydellote's British friends. One English wartime correspondent confided to Aydellote (in a private and confidential letter) that he saw America as the successor to Britain in bringing "English social and political ideals" to the world. Moreover, as long as America won the war by championing these ideals, "I shall feel just as much pride and pleasure in her triumph as in a triumph for these islands."[40]

What we think of as the "standard" curriculum is anything but apolitical. It was created for a particular political purpose and was framed to reflect particular national aims. The multicultural curriculum can be seen as attempting the same type of reform. But whereas the reformers of 1918 had only to deal with European countries, our world has enlarged such that other nations must also be included. However, no one knows what "World Culture(s)" is or should be. What is to be put in? What is to be left out? What counts as a culture? Is there such a thing as a pure culture, a pure tradition? These arguments get reinvented daily in meetings of educational curriculum committees across the country. Meanwhile, nothing exists to replace "Western Civ."

The most obvious successor is "Intro. Bio." The equilibrium within the body politic between the physical and social bodies is changing toward the study of the physical body. Whether one agrees or disagrees with sociobiology and the Human Genome Project, these are the artifacts of an academic society wherein biological explanations are superseding social explanations of culture.

There are several reasons for the ascendancy of biology. First, biology (as a discipline) still maintains that there are truths to be discovered by the creative interaction of human minds with nature. The humanities have adopted the anarchy of postmodernism and have given up the notion of truth; physics has relinquished its leadership of the sciences after quantum probability theory told it there was no way of determining truth; the social sciences have fragmented in a massive identity crisis; biology, in contrast, has become vigorous, multidisciplinary, and relatively well funded. Its reliance on living matter has prevented it from going the route of physics, and its existence within a country suspicious of evolution has kept it from embracing postmodernism. It cannot afford to say that it does not have a more valid, truth-seeking program than that of the creationists.[41] Biology thus salvages one of the most fundamental components of the "Western Civ" tradition, the discovery of truth.

Second, biology deals with the social issues of our day. Its research informs our discussions of race, gender, class, ethnicity, health, food production, industrialization, land use, birth control, law, and education. (Not that biology necessarily *should* inform discussions in all these areas, but it certainly *does*.) Knowing biology is becoming essential for our social discourse. With the demise of the "Western Civ" tradition and with nothing to replace it, biology is taking over its role in the construction of our social judgments. Moreover, at a time when people will not speak of cultures, they can see biology as underlying them all. If part of the value of the Western Civ course was to provide an informed electorate, "Intro. Bio" is taking on this role.

Third, biology is where the action is, and it's where the money is. Every area of biology is expanding, whether it be molecular biology, medical research, or ecology. Biology departments across the country are experiencing enormous increases in majors. There are numerous reasons for this. It is an incredibly exciting time to be a biologist, and the phenomena of life are intrinsically fascinating. The National Science Foundation's push to make science (especially biology) more fashionable in the secondary schools has also worked. Biology is a growth industry, and capital has moved into it, whether it be molecular biotechnology, computer-aided prosthetics manufacture, rational drug design, transgenic crops, or environmental monitoring systems. What is more important than health, food production, and the environment? The ascendancy of biology is being fueled by important social concerns.

Fourth, biology is beginning to provide models for society. This is nothing new. The state-as-an-organism model is implicit in the body politic metaphor, and biologists as varied as Spencer, Hertwig, Just, Goldschmidt, Metchnikov, and Virchow have seen the body as the proper model for society. However, biology has never had much power in the curriculum, and its role in producing explanatory stories for society was not taken very seriously. This has changed. Biological stories are becoming increasingly important in our construction of legal, educational, and health systems. The rhetoric of the Human Genome Project and sociobiology has fashioned stories of culture being genetically defined, and the effect is snowballing. Already, departments of psychiatry no longer focus on consciousness but on neuronal receptors. Education is seen as being wasted on those whose minds are not programmed to receive the teaching.[42] What is legal or illegal, right or wrong, is being redefined in terms of some ill-defined biological imperatives. The "differences" between males and females, blacks and whites, upper class and lower class, European and Asian, are seen as being genetically determined and largely indifferent to social modification. There are other stories coming from other parts of biology (epigenetic interactions from embryology;[43] homeostatic interactions from ecology;[44] new origin stories from primatology).[45] What they have in common is using biology to create stories that define our culture. What had been provided by Greek mythology, the Bible, Dante, Shakespeare, Rousseau, and Goethe is being provided by our interpretations of DNA, cells, organs, animals, plants, and ecosystems.[46] If biology is going to be used to create such stories, the history and critique of biology are more needed now than ever.

We are returning to Aesop, making parables from our study of animals and their parts. But are these stories valid or one-sided? Many of the ones told today certainly are one-sided, and this is why critiques of biology are important and why scientists should know them! Turbulent times (and science and technology add to the turbulence) will demand an order. "Western Civ" is not there to order them, and no substitute has worked. I would argue that biology is going to provide the order that society has been unable to provide. Although I have great confidence that biology can provide meaningful, beautiful, and robust stories that could foster harmony, cooperation, and mutual respect among peoples, it can also produce (and has produced) some of the most pernicious and hateful narratives of human history. Periods of disorder have provided the preconditions for totalitarian narratives and governments. All of a sudden, biology and its historians have more responsibility than they ever wanted.

Notes

1. This essay originated from discussions in the Committee for Educational Policy, Swarthmore College, 1993–94, and I wish especially to acknowledge Drs. Miguel Diaz-Barriga and Nathalie Anderson for their research and conversations on curriculum reform. I also wish to thank editors Peter Taylor, Saul Halfon, Paul Edwards, and Donna Haraway for their comments, suggestions, and questions.
2. Jonathan Parry, "The End of the Body," in *Fragments for a History of the Human Body, Part 2, Zone 4,* ed. Michel Feher (New York: Zone Press, 1989), 491–517; Emily Martin, "The End of the Body?"*American Ethnologist* 19 (1992): 121–40.
3. Scott F. Gilbert, "Resurrecting the Body: Has Postmodernism Had Any Effect on Biology?" *Science in Context* 8 (1995): 563–77.
4. George Lakoff and Mark Johnson, *Metaphors We Live By* (Chicago: University of Chicago Press, 1980).
5. Interestingly, the grafts that have the best chance of success are those of brain tissues (and sperm-forming tissue), since these regions are protected from the immune system by extracellular matrices.
6. Jan Klein, *Immunology: The Science of Self-Nonself Discrimination* (New York: John Wiley, 1982).
7. Donna Haraway, "The Biopolitics of Postmodern Bodies: Determinations of Self in Immune System Discourse," *differences: a journal of feminist cultural studies* 1 (1989): 3–43.
8. Scott F. Gilbert, "The Metaphorical Structuring of Social Perceptions," *Soundings* 62 (1979): 166–86.
9. Susan Sontag, *Illness as Metaphor* (New York: Anchor Books, 1979).
10. The first popular magazine articles on Alzheimer's disease came out in 1977; the first on AIDS came out in 1981.
11. Paula A. Treichler, "AIDS, Homophobia, and Biomedical Discourse: An Epidemic of Signification," *Cultural Studies* 1 (1987): 263–305; Sander Gilman, "AIDS and Syphilis: The Iconography of Disease," in *AIDS: Cultural Analysis, Cultural Activism,* ed. Douglas Crimp (Cambridge: MIT Press, 1988), 87–107.
12. David Baltimore, "The Brain of a Cell," *Science* 84, no. 11 (1984): 149–51.
13. Dorothy Nelkin and M. Susan Lindee, *The DNA Mystique: The Gene as a Cultural Icon* (New York: W. H. Freeman, 1995).
14. Oscar Hertwig, *The Biological Problem of To-day: Preformation or Epigenesis? The Basis of a Theory of Organic Development,* trans. P. Chalmers Mitchell (New York: Macmillan, 1984), 135–36; Alfred I. Tauber, "From Self to the Other," in *Metamedical Ethics: The Philosophical Foundation of Bioethics,* ed. Michael A. Grodin (Dordrecht: Kluwer Press, 1995); Alfred I. Tauber, *The Immune Self: Theory or Metaphor?* (New York: Cambridge University Press, 1994); Scott F. Gilbert and Steven Borish, "How Cells Learn: Induction, Competence, and Education within the Body," in *Change*

and Development: Issues of Theory, Method, and Application, ed. K. Ann Reninger and Eric Amsel (Hillsdale, N.J.: Lawrence Erlbaum, 1997).

15. Scott F. Gilbert, "Adaptive Enzymes and the Molecularization of Embryology," in *The History of Molecular Biology*, ed. Sahotra Sarkar (Dordrecht: Kluwer Press, 1996); Scott F. Gilbert, John M. Opitz, and Rudolf A. Raff, "Reintegrating Developmental and Evolutionary Biology," *Developmental Biology* 173 (1996): 357–72.
16. See Tauber, "From Self to the Other"; Gilbert, Opitz, and Raff, "Reintegrating Developmental and Evolutionary Biology"; Nils K. Jerne, "The Generative Grammar of the Immune System," *EMBO Journal* 4 (1985): 847–52; Brian Goodwin, *How the Leopard Changed Its Spots: The Evolution of Complexity* (New York: Scribner's, 1994); Francisco J. Varela and Antonio Coutino, "Second Generation Immune Networks," *Immunology Today* 12 (1991): 159–66.
17. Sandra Harding, *Whose Science? Whose Knowledge? Thinking from Women's Lives* (Ithaca, N.Y.: Cornell University Press, 1991), 10. Stephen J. Gould has written about science in a similar vein in "Shields of Expectation—and Actuality," in *Eight Little Piggies* (New York: Norton, 1993), 409–26.
18. Eugene Marion Klaaren, *The Religious Origins of Modern Science* (Grand Rapids, Mich.: W. B. Eerdmans, 1977).
19. Harding, *Whose Science?*; and Sandra Harding, ed., *The "Racial" Economy of Science: Toward a Democratic Future* (Bloomington: Indiana University Press, 1993).
20. Sal Restivo, "Modern Science as a Social Problem," *Social Problems* 35 (1988): 3.
21. Ernst H. Haeckel, *The Riddle of the Universe* (New York: Harper, 1902); Madison Grant, *The Passing of the Great Race; or, The Racial Basis of European History* (New York: Scribner's, 1916); Karl C. Vogt, *Lectures on Man* (London: Longman, Green, Longman, and Roberts, 1864). See Stephen J. Gould, *The Mismeasure of Man* (New York: Norton, 1981).
22. Roger V. Short, "Sex Determination and Differentiation," in *Reproduction in Mammals: Embryonic and Fetal Development*, ed. Colin R. Austin and Roger V. Short (Cambridge: Cambridge University Press, 1972), 70. This passage and others like it are not to be found in the 1982 revision of this textbook. This example was used in Bonnie Spanier, "The Natural Sciences: Casting a Critical Eye on 'Objectivity,'" in *Toward a Balanced Curriculum*, ed. Bonnie Spanier, Alexander Bloom, and Darlene Boroviak (Cambridge: Schenkman Publishing, 1984), 49–57.
23. Biology and Gender Study Group, "The Importance of Feminist Critique for Contemporary Cell Biology," *Hypatia* 3 (1988): 61–76; Emily Martin, "The Egg and the Sperm: How Science Has Constructed a Romance Based on Stereotypical Male-Female Roles," *Signs* 16 (1991): 485–501; Meredith F. Small, "Sperm Wars: The Battle for Conception," *Discover* (July 1991): 48–53.
24. Scott F. Gilbert, "Cells in Search of Community: Critiques of Weismanism and Selectable Units in Ontogeny," *Biology and Philosophy* 7 (1992): 473–87; Evelyn F. Keller, *Secrets of Life, Secrets of Death* (New York: Routledge, 1992).
25. See, for instance, quotations of Rabi in John S. Rigden, *Rabi: Scientist and Citizen* (New York: Basic Books, 1987). On the importance of this idea in the United States, see David A. Hollinger, "Free Enterprise and Free Inquiry: The Emergence of Laissez-Faire Communitarianism in the Ideology of Science in the United States," in *Science, Jews, and Secularity: Studies in Mid-Twentieth-Century American Intellectual History* (Princeton, N.J.: Princeton University Press, 1996).
26. Alvin Toffler and Heidi Toffler, *Creating a New Civilization* (Atlanta: Turner Publishing, 1995).
27. Cinna Lomnitz, *Fundamentals of Earthquake Prediction* (New York: John Wiley, 1993).
28. Harding, *Whose Science?* 33.

29. Thomas H. Huxley, *Science and Education* (New York: Appleton, 1897).
30. Edward Everett, quoted in Eric Temple Bell, *The Queen of the Sciences* (Baltimore: William and Wilkins, 1931), 20.
31. Oswald Spengler, "The Meaning of Numbers," in *The World of Mathematics*, ed. James R. Newman (New York: Simon and Schuster, 1955), 2315–47.
32. Robert N. Proctor, *Value-Free Science?* (Cambridge: Harvard University Press, 1991).
33. See, for example, Ruth Bleier, *Gender and Science* (New York: Pergamon Press, 1977); Anne Fausto-Sterling, *Myths of Gender* (New York: Basic Books, 1985); Donna J. Haraway, *Primate Visions: Gender, Race, and Nature in the World of Modern Science* (New York: Routledge, 1989).
34. Gilbert, "Cells in Search of Community"; Daniel P. Todes, *Darwin without Malthus: The Struggle for Existence in Russian Evolutionary Thought* (New York: Oxford University Press, 1989).
35. Gilbert Allardyce, "The Rise and Fall of the Western Civilization Course," *American Historical Review* 87 (1982): 695–725.
36. Ibid.
37. Milledge Louis Bonham, "The War Issues Course," *Southern School 1919*: 346–47. In Aydelotte Archives Box 99, Swarthmore College.
38. Frank Aydellote, letter to Dr. Morris P. Tilney, University of Michigan, 1921. Ayedelotte Archives Box 99, Swarthmore College.
39. Frank Aydellote, letter to Dr. C. R. Mann, War Department Committee on Education, 1918. Ayedelotte Archives Box 99, Swarthmore College.
40. Confidential letter to Frank Aydellote. Ayedelotte Archives, Swarthmore College.
41. See Gilbert, "Resurrecting the Body."
42. Richard J. Herrnstein and Charles Murray, *The Bell Curve: Intelligence and Class Structure in America* (New York: Free Press, 1994); J. Philippe Rushton, *Race, Evolution, and Behavior: A Life History Perspective* (New York: Transactions Press, 1994).
43. Gilbert and Borish, "How Cells Learn."
44. Lynn Margulis, *Symbiosis in Cell Evolution* (San Francisco: W. H. Freeman, 1981); Lewis Thomas, *The Lives of a Cell: Notes of a Biology Watcher* (New York: Viking Press, 1974).
45. Haraway, *Primate Visions*.
46. Indeed, the television programs starting from Disney nature programs and continuing through *Nova* and *Discovery* resemble the Moody Bible Institute's "Sermons from Science" series. So let me end by telling a tale, extrapolating a biological concept into discussions of society and the multicultural curriculum. *Homology* is a word that comes from comparative anatomy and embryology. Although this word means somewhat different things to different groups of biologists, it generally implies that two or more structures (bones, organs, proteins, nucleic acid sequences) have an underlying similarity despite their obvious differences. For instance, here are three sets of homologous structures: (1) our index finger and any other finger; (2) any finger and any toe; (3) the human hand and the bird's wing. So what does one emphasize, the differences between our toes and our fingers or their similarities? Nobody will deny that they are different. Nobody will deny that they are similar. Whether one emphasizes the differences or the similarities between homologous structures is thus a political or an aesthetic decision. Does one stress the differences between men and women, between Chinese cultures and Western cultures, between Serbs and Croats, between feather development and hair development, or does one emphasize their similarities? In this manner, I am homologous to any other person in the world. No person is totally "other," "foreign," or "exotic," nor is any person the same as me. Homology diffuses "the exoticism of the other" while at the same time allowing us to celebrate both our

real similarities and our real differences. Homologies need not even serve the same functions (i.e., they are not necessarily analogous). The bones we use for hearing are homologous to the bones fish use for chewing. By being aware of both the differences and the similarities between groups, we can expand the liberal arts repertoire and teach students to recognize what is excellent in a wider cultural context.

Genetic Engineering, Discourses of Deficiency, and the New Politics of Population

Herbert Gottweis

Questions of Identity

"Who is us?" asks Robert B. Reich, former political economy lecturer at Harvard and U.S. Secretary of Labor in the first Clinton administration, in a 1990 issue of the *Harvard Business Review*.[1] "Who is 'us'? Is it IBM, Motorola, Whirlpool, and General Motors? Or is it Sony, Thomson, Philips, and Honda?"[2] And, after a discussion of the relationship between national interest and increasingly globally acting industry, Reich comes to the stunning conclusion: "The answer is, the American workforce, the American people, but not particularly the American corporation."[3]

Relieved that "we" are neither "Whirlpool" nor "Honda," we can go on to appreciate the deeper meanings of Reich's pondering. "Who are we?": this question refers to the spatiotemporal conditions of democratic practice. Reich's reading of the contours of American collective identity points to the dilemmas of stabilizing collective self-images in times of turbulent changes in the global political economy. As it becomes increasingly difficult to draw lines of collective identity along geographical boundaries, these identities are culturally rewritten in terms of a new domestic/global narrative. With the gradual breakdown since the 1970s of the postwar order, the rise of transnational capitalism, and the breakdown of the regimes in Eastern and Central Europe and in the Soviet Union, a new discursive space has opened up for recasting the architecture of the world order.

Identity is established in relation to differences that have become socially recognized.[4] Many would identify IBM as an American and Honda as a Japanese company. But what if both IBM and Honda act increasingly transnationally rather than nationally? Or, to give another example: is the pharmaceutical company Hoffmann-La Roche a Swiss, a British, an American, or a Japanese company? Roche's Central Research Unit is based in Basel; but the Roche Institute of Molecular Biology is located in Nutley, New Jersey; important sites for product development are in England and Japan. Does it, then, make sense to view IBM as American and Hoffmann-La Roche as Swiss?[5]

Reich's answer is no. He redefines American collective identity by resorting to the workforce as anchor of stability. Although capital may be hypermobile and transnational, workers are not. Hence, for him it is the workers who constitute valuable objects of investment and points of departure for redefining America.

Reich's rewriting of contemporary U.S. collective identity is a response to the new power of transnational corporations that proposes the ideology of productivism as the source of rewriting "us." We produce, therefore we are. But underlying this ideology of the "we" is the liberal version of the subject that tacitly enters Reich's equation as citizen, rational human being, voter, person.

Who are we? Human Genome projects, today pursued simultaneously in a variety of countries, offer different answers. The European Community (EC) gives the following interpretation: "The human genome is the complete set of genetic material . . . which embodies the instructions describing each human being. It is now within the realms of possibility to 'read' these instructions in their entirety."[6] We are our genes. Eureka!

Both of these answers to the question of who we are inscribe subjectivity/collectivity into the sociopolitical space. Not that the one definition follows from the other. But, as I argue in this essay, they intersect at various points as narratives of life and narratives of production and exist in relationships of mutual reinforcement. Neither "we" as a collective, nor the "state," nor "life" are "stable entities." What counts at a particular point in time as knowledge of "the economy" or of "life" is an effect of fixation, of a stabilization of differences. These systems of belief are upheld by regimes of truth, affect the ways we understand ourselves, and constitute, as discursive economies or epistemic configurations, a crucial dimension in the production of the social.

The idea of what we are has been challenged by the discourse of modern biology over the last decades. As genetic engineering promises to offer new, powerful means to "rewrite life," these attempts at rewriting have effectively questioned notions of individuality as well as of collective identity. But, so my argument goes, the representational shift from the control of genetic molecules to the control of the genetic sequences is creating new forms of social disciplining and policing: the new genetic technologies play a crucial role in shaping a new matrix of surveillance and intervention, adding to the recasting of contemporary panopticism.[7] At the same time, critics have identified molecular biology's systems of signification as strategies of cultural infiltration, which change and normalize parameters of human self-recognition and the understanding of nature. Social movements began to counter these strategies by a politics that would allow a space for alter identities.

My essay highlights this contested process of rewriting subjectivity/us by means of genetic technologies. Political identity originates through processes of reinscription in multiple locations. Thus, the discursive practices of modern genetics are inseparable from and related to processes of rewriting in other sites, such as industrial production, modes of interaction between state and civil society, and tendencies in the configuration of the capitalist world order. It is the genealogy of this relationship between identity/difference, narratives of science, production, and the political that I want to explore. First I will show how the discourse of molecular biology developed in mutual reinforcement with the rise of multinational liberalism and the shaping of American hegemony in the capitalist world system. The subsequent discussion underscores how the hegemonic crisis of the post-World War II order in the 1970s reinforced the interventionist potentials of molecular biology in the form of genetic technologies. I will demonstrate the emergence of a new, discursive economy, a new sociopolitical space, constitutive of a new system of interrelations: a configuration consisting of molecular biology's representations of life, efforts to construct a new economic space in the form of the biotechnology industry, the reshaping of parameters of human self-recognition through new modes of representing life, and the construction of a new matrix of surveillance. I conclude that these strategies in the politics of identity/difference to stabilize a new sociopolitical space, which were built around the social powers of molecular biology, remain closely tied to ongoing contestations by social movements resisting such efforts in the consolidation of the architecture of political order. Furthermore, there is mounting evidence that some of the core assumptions made in the research tradition in molecular biology are being deconstructed by events in research and product development practices and in biological discourse.

The Government of Populations and Molecules

Who are we? As already established, subjectivity emerges out of sets of differences that are socially recognized and coded. Hence, subjectivity and self are social constructs that can be understood in terms of their historicity. Michel Foucault has pointed to a significant historical transition contemporaneous with the emergence of industrial capitalism, in which a shift of emphasis occurred from the primacy of sovereignty, law, and coercion or force "to take life" to the emergence of new forms of power constitutive of life.[8] This power over life evolved in two forms: disciplining the body and regulating populations. Whereas the former had as its object the individual, the latter addressed itself explicitly to the "ensemble of the population" as a field of shaping and forging.[9] These two strategies constituted the two poles around which the power over life was organized.[10]

With the rise of molecular biology in the 1930s, modes of representing life have changed significantly. The Rockefeller Foundation in particular, in alliance with a group of leading scientists, was the driving force in shaping a "new science of man" that was to provide new levels of social control by increasing power over life processes. In 1933, the Rockefeller Foundation had inaugurated the new program that in 1938 came to be called "molecular biology." According to the foundation, the previous one hundred years had marked the supremacy of physics and chemistry, to the neglect of questions of human behavior. This new goal was to attain social control through scientific understanding. Warren Weaver, the program director, compared the divisibility of the cell, the older unit of analysis, into subcellular units to the divisibility of atoms to subatomic units. According to Lily E. Kay,

> this molecularization of life was to be applied to the study of genetic and epigenetic aspects of human behavior. Based on the faith in the power of upward causation to explain life, Weaver and his colleagues saw the program as the surest foundation for a fundamental understanding of the human soma and psyche, and ultimately the path to rational social control.[11]

The new selected fields were targeted by the foundation for enormous grants. The funding was not restricted to the United States, but operated on a global scale, with leading European scientists receiving major funding. The Rockefeller Foundation's promotion of the new biology was closely related to the decline of old-style eugenics. At a time when the "old," "crude," "population-oriented" eugenics became increasingly scientifically and politically unacceptable, the much more subtle, subcellular method to control life became "the technology of choice."[12]

Since the 1930s, the discourse of molecular biology has attempted to explain biological function exhaustively through the study of physical processes and chemical structures. Biological phenomena were conceptualized as the consequences of the interaction of large assemblies of biologically significant molecules and it was claimed that the constitutive chemical components determined, and ultimately explained, the foundation of all biological phenomena. Properties of life, health, and disease were defined in terms of genetically directed macromolecules. A new space of representation had developed providing a new language and system of signification of life. However, there was a critical link of continuity between eugenics and molecular biology. The object of intervention had changed from populations (gene pool) to the subcellular level, but geneticization—the framing of central dimensions of humans in terms of genetic factors—continued to inform scientific practice. The genetic determinism of eugenics had given way to the molecular reductionism of the new biology, but the control of life processes on the genetic level remained the goal.

The Rockefeller Foundation's engagement in the "new science of man" must be understood against the background of dramatic changes in the global political economy. With World War I, the Eurocentric nineteenth-century order had come to an end. The international configuration of forces took the form of spheres of influence and rival imperialist blocs.[13] In October 1929, after the Wall Street crash, a general economic decline took hold. By 1932, many currencies were floating and international finance had virtually collapsed. At this time of crisis in the United States, a new historical bloc emerged. Capital-intensive industries, investment banks, and internationally oriented commercial banks became the key elements in this new configuration. This new hegemonic bloc was at the core of the New Deal, with its ability to accommodate millions of mobilized workers during the Depression. The capital-intensive firms used less direct human labor, felt less threatened by labor turbulences, and had the space and resources to accommodate the workers. In addition, being world and domestic leaders in their fields, they could only gain from free trade.[14] The Rockefeller Foundation's engagement in social reform was congruent with the emerging new system of power relations. On the economic level, the projects in the social and biological sciences were intended to increase economic productivity and social stability worldwide; at the same time, motives of social control and surveillance directed the foundation's interest in the human body, on the individual as well as on the aggregated level.[15] From its inception, the new hegemonic project of multinational liberalism involved not only political economy, but a politics of life ultimately constitutive of its rationality. This process was not propelled by any monolithic institution confronting individuals from above.[16] Rather, the process of change was introduced by a power matrix of philanthropic institutions such as the Rockefeller Foundation, public and private universities, governments, and groups of scientists building on strategies for a new knowledge of life. Unlike the much more state-power-dependent eugenic movement, molecular biology operated in a rather decentralized way around various nodal points in an emerging hegemonic formation. Molecular biology, hence, developed in mutual reinforcement with the rise of multinational liberalism, which, in the form of the Pax Americana, would soon become the dominant ordering principle of the post-World War II economic order.

The new U.S.-centered postwar order developed along several interrelated dimensions: the political, economic, and sociocultural (re)construction of the defeated Axis powers; the economic reconstruction of Western Europe under the Marshall Plan; the militarization of U.S.-European relations through NATO; and the waging of the Cold War, primarily directed against the Soviet Union.[17] The central premise of the new hegemonic order, a liberal international system, was to establish the unrestricted globalization of the structure of accumulation that had emerged at the end of the nineteenth century.[18] The neoliberal state operated based on the system of corporatist government-business-labor coordination. On the social side, the liberal international system was supported by the "internationalization of the New Deal" with a promotion of the social and the state protecting vulnerable social groups. But the new political discourse also framed the state as an agency with significantly more responsibilities on the economic side. One important field of increased state interventionism was science and technology policy. The political-economic discourse in the post-World War II era defined scientific development and economic growth as intrinsically linked to each other.

As a result, governmental research-and-development spending grew worldwide at unprecedented levels. This increase was partially driven by the argument that the state had to support industry by engaging in research; and it was partially influenced by dramatically increased defense expenditures, in the United States as well as in Europe. Research and development became a core element of the evolving politics of security. This politics of security focused on the "ensemble of the population," in its outward as well as its inward orientation. The discourse of foreign policy, with the Communist menace to the collective as its point of orientation, was part of a governmental rationality of which the promotion of the social—the well-being, health, and prosperity of the population—were other crucial elements.[19] In this perspective, an important aspect of the Pax Americana was the emergence of a particular kind of individual and collective identity. This occurred by creating social cohesion through various interrelated practices and methods, such as military policy, industrial policy, and welfare provisions.

After the Second World War, Western Europe, Japan, and the United States rebuilt their economies after a twenty-year period of depressed consumer demand. Between 1945 and 1973, favorable conditions in the United States stimulated high growth rates throughout the world economy. The growth of the U.S. economy helped Europe and Japan to reconstruct their own war-devastated economies. The reconstruction was based on American mass-assembly principles and was designed to achieve the "economic miracles" of the late 1950s. Pent-up consumer demand for homes, household appliances, and other consumer goods stimulated increased production. Technological innovations in electric equipment, machinery, scientific instruments, transportation, and in chemical and space vehicle industries caused rapid growth. All these conditions increased total demand, resulting in rapid economic growth.[20] The world economy was geared toward a never-ending boom that, among other things, seemed to be linked to scientific progress.

In the decades after World War II, molecular biology was increasingly abandoned by the Rockefeller Foundation, and state support gained in importance. A crucial part in the writing of any policy consists in determining an object's significance within a larger context, that is, in establishing the meaning of an object in the interdiscursive constellation of politics, where various political discourses, such as the discourse on welfare, economic development, and foreign policy, intersect and constitute the text of politics. To understand how any sectoral policy evolves, it is crucial to situate the policy within the broader context of images, ideas, representations, and social projects genuine to a society.

As already mentioned, the representation of scientific and technological progress as a central condition for any nation's viability was a core feature of the politics of modernization and gained prominence across industrialized nations after World War II. As we look at the ways policies of molecular biology were shaped in countries such as the United States, Britain, Germany, and France, we see an evolving tactic of representing molecular biology as intimately tied to the general social project of modernization, civility, and economic growth. This discourse of science was related to a more general shift in the political-economic discourse on the significance of science and technology for economic development. In the post-World War II era, the political-economic discourse in the industrialized countries defined scientific development and economic growth as intrinsically linked to each other. The relationship was explained via the strong correlation between dramatically increased research-and-development spending after the Second World War and the period of unprecedented economic growth stretching into the early 1970s. During the 1960s, the "science push theory," which argued that the key to innovation was the "push" of scientific development, gained in importance; and it was the state that had to push science.[21] Especially during the 1960s and 1970s, science and technology policies were being systematically developed in the industrialized countries. Scientific and technological progress became increasingly represented as a central condition for a nation's viability as expressed by economic growth, agricultural expansion, and medical advances and its capability to compete in the international economy.[22]

Within this discursive economy, molecular biology mutated from its interpretation as a small, avant-garde field to a "futures technology" with potential to alleviate some of humanity's main scourges, such as disease and hunger. At the time, the argument was that strong support for science would, sometime and somehow, translate into social and economic benefit. Molecular biology came to be represented as related to the general dominant social project of modernization. At the same time, the articulation of molecular biology into the activity of government was coevolving with a discourse that coded modernization as accelerated industrialization (signified by continuous economic growth). The framing of molecular biology as a moment in the politics of modernization established a relationship between the inscription of modernity qua industrialization and the emergence of a sectoral policy of molecular biology. Although molecular biology in the 1950s and 1960s was generally considered "basic science," this view obscures its political-economic coding. Soon a politics of molecular biology emerged that constructed scenarios of challenges and threats to society that required the state to engage in "policies of salvation" and displayed the state's powers of intervention.

Institutionally, this discourse was first stabilized in the United States with the dramatic growth of the National Institutes of Health (NIH), which was about to become the main state agency to support molecular biology. In fact, since the mid-1950s, the NIH heralded the rise of state-dominated biomedicine in the United States, thus completing the Rockefeller Foundation's never fully successful attempts at a comprehensive support of the biomedical sciences.[23] What was accomplished over the years was the creation of a highly specialized cadre of experts grouped together in universities, medical schools, the various branches of the NIH, and clinical medicine.[24]

During the 1950s and 1960s, the major U.S. investments in molecular biology, and the resulting developments in the field, increasingly caused concern in Europe over a "delay" and the "backwardness" of European research. These concerns translated into various national and European initiatives to deliberately promote the new field. In the United Kingdom, the Medical Research Council (MRC) emerged as the major sponsor of molecular biology;[25] in France, the Délégation Générale pour la Recherche Scientifique et Technique (DGRST) selected molecular biology as a special field of state support; and in Germany, the German Research Foundation (DFG), the Max-Planck-Society, and the public/private hybrid Volkswagen Foundation identified biology as a field of considerable potential for the future. A representative of the Volkswagen Foundation recalled the situation: "We wanted . . . to build up modern biology as quickly as possible. . . . What the Rockefeller Foundation had anticipated and simulated in America, we could see there and we wanted it for Germany."[26] The creation of the European Molecular Biology Organization (EMBO) signified visions on the international level. The idea behind EMBO was to promote a concentration of European efforts in the new biology by encouraging training, mobility, and communication by providing an adequate infrastructure and special laboratory facilities, a setup in many ways reminiscent of the Rockefeller Foundation's strategies since the 1930s.[27]

With clusters of programs and support measures requiring the cooperation of scientists, administrators, and politicians on the national as well as on the international level, the field of biology changed significantly. What had emerged was the transnational space of molecular biology, defined by an ongoing stream of meetings, collaborations, and correspondence between scientists and administrators, mainly in the United States and Europe.[28] Even before specific techniques became available in the 1960s, the term "genetic engineering" gained currency, denoting the deliberate and controlled modification of genetic material. Molecular biology became increasingly construed as a new and important field, revealing not only the secrets of life but also having enormous potential economic and medical implications, and thus deserving close attention from the state. For example, a report of a 1966 "working group" set up by the British Council for Science Policy and chaired by John Kendrew came to the conclusion that progress in biology took place chiefly in the area of molecular biology. Not only did the impact of research in molecular biology have "deep intellectual significance" but it was likely to yield "social and economic dividends of inestimable value" through biomedical and agricultural applications.

During this time, a public image of molecular biology was produced, portraying it as the new, "hot" direction in the life sciences. Scientists, journalists, philosophers, and theologians strung a more or less continuous chain of interpretations around the meanings and implications of the new biology. In general, scientists tended to portray molecular biology as a new, revolutionary field, while the media responded with some concerns about the potential abuse of the new biology. Already in the 1960s voiced warnings about the possible negative cultural and political impact of the new biology received much attention in the media. In particular, the 1962 Ciba Foundation conference "Man and His Future"—a gathering of a scientific elite in genetics, evolutionary biology, medicine, and biochemistry—and the publication of its proceedings provoked something of an international uproar. The participants agreed that the "new biology" would finally allow humankind to master evolution. Science would offer the means to do so. Direct manipulation of germ lines and the deliberate support of positive heredity traits were elements of a broader strategy designed to establish a better future for humanity.[29] These proceedings created strong and critical responses all over Europe and the United States. For many, the new biology's images of "man and his future" evoked associations of horror rather than hope.

The Emerging Discourse of Deficiency

Nurtured by massive state support, molecular biology continued to develop, and the United States became its intellectual center. By the mid-1970s, all the techniques necessary for genetic engineering with bacteria were available. A dramatic shift in molecular biology seemed to be in the air. Whereas biochemistry and early molecular biology had attempted to produce an extracellular space of representation for intracellular structures and processes, genetic engineering strove to do the reverse: to create, by means of information molecules, the intracellular representation of an extracellular project. In other words, genetic engineering began to "rewrite" life.[30] The interventionist potentials, which had been inscribed into molecular biology from the very beginning, turned into veritable instruments of transforming the structure of life on the molecular level. The discourse of molecular biology began to contribute to what could be called a "discourse of deficiency," the rewriting of life on a subcellular level in terms of "absences," of "improvables" in need of the intervention of genetic technologies. The biological discourse inscribed bacteria, animals, plants, and human beings as potentially in a state of deficiency and in need of supplemental genetic technology.

Genetic engineering arrived at a time that also marked the beginning of a hegemonic crisis in the postwar order and of the centrality of the United States within the architecture of the Pax Americana. The promise of the new technology, so diligently promoted by scientists since the 1960s, met with the frantic search of governments for ways out of the deepest recession since the 1930s. With the oil crisis, surging inflation, and the economic recession in the first half of the 1970s, the dream of the never-ending post-World War II economic boom ended abruptly. The era of cheap energy was over and the newly industrializing countries posed a new challenge to the established players in the global economic system. Furthermore, the rise of transnational companies brought about a shift of power between the national and the transnational levels: finance and money became increasingly decoupled from the system of production and trade, causing major contradictions for economic development.[31] With these developments the postwar capitalist growth model came increasingly under attack from neoliberal and conservative critics. They represented the postwar model as containing a set of barriers, such as powerful trade unions, state monopolies, and lack of investment incentives, which together inhibited market dynamics and kept national economies from adapting and facing up to new opportunities and challenges. Only by removing these barriers could recovery be expected. Elements of this "strategy of recovery," enacted in various policy fields, included increasing the profitability of private capital; creating a more favorable labor market by dismantling trade unions; reducing corporate and middle-class taxation; and privatizing public industries and banks.[32]

The trope of the "international high-tech race" between the United States, Japan, and Europe gained currency. Successful participation in this competition, so the argument went, was tied to a removal of national obstacles to innovation. Hence, technology policies, trade policies, and industrial policies acted as mutually reinforcing discursive practices in a process of reshaping the architecture of the political order. As individual countries increasingly lost the capacity to successfully steer national/autonomous policies, the political focus shifted to the creation of economic, regulatory, and political frameworks attractive for the newly emerging forces in the global economy.[33] The post-World War II social compromise with government/business/labor cooperation at its core was abandoned, with labor being pushed into a defensive posture. The more confrontational version of this process was realized during the Reagan and Thatcher administrations, while a more consensual approach prevailed in countries such as Japan and West Germany.[34] At the same time, the autonomy of the state was challenged by a surge of extraparliamentary political mobilization in the form of new social movements; for instance, in Europe, Green parties emerged to question the contours of the post-World War II political order.

Hence, the crisis of the 1970s was more than a deep recession. It was the onset of a deep transformation in the domestic systems of social regulation and in the global political economy. New information technologies and biotechnology were framed as the driving forces in this transition. The appreciation of the new technologies was not limited to their economic value, but also extended to their potential to exert social control. An example of this perspective can be found in a 1980 report of the European Commission's Technology Forecasting Office, FAST, which was commissioned to evaluate the need for a Community strategy in biotechnology. The authors of this document came to the following evaluation: Europe is in a process of increased uncertainty concerning future social, economic, and political developments. Science and technology could play a key role in mastering this process of transition:

> Though labor troubles, unemployment and energy currently dominate the international economic scene, the coming twenty to thirty years will, it is thought, see two major changes: the computerization of society [and] . . . the biological revolution emanating from the boom of "life technologies." . . . Within the relatively near future, bio-technology could be used in a number of sectors such as: Human health and behavior: we could control the development of the human embryo, and, perhaps within twenty years, determine its sex. We could prevent certain malfunctions. We should be able to create new vaccines and inoffensive drugs to counter addiction to alcohol or tobacco—even to regulate mood and emotions. We could also improve the quality of life for the elderly, improve techniques for transplants.[35]

Clearly, this interpretation reflects a view of the present condition of contemporary society as troubled and "out of control." In addition to yielding economic benefits, genetic engineering was conceptualized as a potential contribution to a broader social stabilization, mainly by virtue of its expected capacity to control behavior and bodies.

This combined view of biotechnology as economic and social strategy was not a singular event. What in 1980 took the shape of informed speculation by policymakers, ten years later reappeared in the form of the Human Genome Project as a better-defined technopolitical strategy. By the late 1980s, the United States, Japan, the Soviet Union, Denmark, France, Germany, Italy, the United Kingdom, and the European Community had launched formal human genome programs.[36] As Leroy Hood, one of the main proponents of human genome research in the United States put it,

> the human genome project is on its way to creating an encyclopedia of life, giving biologists and physicians direct computer access to the secrets of our chromosomes. Just as the complex road system of the United States has transformed transportation in the country by permitting ready access to virtually any city, street, or house, so will the creation of genetic, physical, and sequence maps greatly facilitate our ability to access interesting genes. . . . computer access to the human genome maps will dramatically alter the practice of biology.[37]

Genetic data gathering, inseparable from the process of genetic intervention, is, to be sure, only one dimension of a broader process in the current shaping of "everyday life" panopticism. In the field of international politics, the intersection of communication intelligence, radar intelligence, telemetry intelligence, and photo intelligence has created a new regime of international relations. Together, these techniques contribute to a normalization of war and peace by technical means: the same satellites that monitor nuclear arms treaties can be used to map pathways for low-level cruise missiles.[38] On the domestic level, information gathering by credit and financial companies, by the state and its various agencies, by insurance and health firms, and by workplace-related surveillance in the form of bar coding interconnect to form a matrix of new forms of social control.[39] Just as military surveillance collapses peace and war, domestic forms of surveillance—such as the construction of DNA data banks, the representation of life in forms of genetic linkage and physical maps—remain inseparable from genetic interventions, such as genetic testing and attendant recommendations.[40] The FBI, for example, is creating such a data bank containing the DNA fingerprints of suspected criminals. Although collected for the purpose of criminal investigation, the stored DNA specimens could, of course, be used for other purposes.[41]

This project finds itself most clearly summed up in the strategic deliberations on a European Community strategy for biotechnology in the term "Bio-Society," coined by the European Commission's FAST unit. The "Bio-Society" is, according to the EC, shorthand for a scenario in which, over the coming decades, significant sections of human activity may be transformed from their traditional forms into forms based on the exploitation of the techniques and products of biotechnology. In other words, it is a scenario for a society whose identity is secured and at the same time redefined by the discursive practices of biotechnology.[42]

In this scenario of a Bio-Society, we most clearly recognize the relationship between new modes of global research and production and the deployment of surveillance strategies: the shaping of a new economic space, the reshaping of parameters of human self-recognition, and the construction of a matrix of surveillance (as articulated in the construction of new systems of DNA information technology) coalesce and link up to a broader picture of the transformation of individuality, political identity, and panopticism. In the new global arena, discipline and production remain tightly interconnected through surveillance, which now takes the form of the new technopanopticism, from the omnipresent observation of social processes by video cameras to the construction of DNA data banks.

As in the 1930s, the idea of forging the subject has been tightly interwoven with broader motives of economic and social development. But, in contrast to the 1930s, postwar developments have ceased to be primarily centered in the United States and instead have operated on a global level in the form of increasingly complex apparatuses of scientific discourses, networks of strategic alliances between companies, scientific organizations, and the state. The state in particular took a decisive role in encouraging the new biotechnologies. Human genome policies are just one example of state intervention in the field of biotechnology. Encompassing policies of regulation and technology wove biology's discourse of deficiency into a strategy of social and economic cohesion and surveillance. After molecular biology had been framed as a "mode of modernization" during the 1960s, now the "new biotechnology" further specified this "project of modernization" in an interdiscursive relationship with industrial policy and trade policy. These policies coded biotechnology as a central "high-technology" field expected to determine national futures.

With these developments, the social space for a democratic negotiation of the introduction of genetic engineering into society had narrowed significantly. The arrival of genetic engineering immediately stirred deep concerns among scientists, as well as the public, about possibly hazardous socioeconomic and cultural implications. Especially in the United States, citizen groups challenged the quick introduction of recombinant DNA techniques without sufficient deliberation and scrutiny. However, this protest remained mostly local and unheard. Following the 1975 Asilomar II conference, which had focused on the hazards of recombinant DNA research, an international consensus emerged among molecular biologists, administrators, politicians, and interest groups from industry and science to go ahead with the research under initially self-imposed regulatory guidelines. The strategy of this consensus was to represent hazard as controlled by the technology that produced it. The adopted regulatory measures essentially combined physical containment (safety measures in the laboratory) with biological containment (choice of bacterial strains as carriers for the recombinant nucleic acid molecules). Thus, the representational strategies dealing with risk connected inseparably with the definition of life within the discourse on life in molecular biology. As the definition of hazard was controlled by the play of signifiers in the discourse of molecular biology, deregulation was simply a matter of the "progress of science."

The 1976 NIH guidelines on recombinant DNA work in laboratories became a legal strategy imitated in most other Western countries. The drive to follow guidelines for work with recombinant DNA was, however, less a matter of conviction than of politics. As the European Molecular Biology Laboratory (EMBL) director remembered:

> Any attempt to establish risks associated with recombinant DNA research is, today, based upon prejudice and conjectures rather than knowledge. Guesswork and intuition rather than objectivity have to be the order of the

> day, and, therefore, it seemed to many Europeans pointless to duplicate the task of assessing risks when it was pursued so energetically in the USA. . . . Moreover, the consensus represented by the draft NIH guidelines seemed to be a workable compromise which had the virtue of internal consistency. These considerations, as well as the recognition of the current hegemony of the USA in this field, made the acceptance in Europe of the NIH draft guidelines a matter of little trauma.[43]

To be sure, none of the other countries adopted the NIH guidelines without revisions or more or less substantial changes. However, overall a great similarity in global regulation had been reached, exemplified by other countries following each of the revisions of the NIH guidelines over the years. From the very beginning, the greatest concern was the existence of a more or less common regulatory framework in order to avoid regulatory distortions in research and development. Governments that did not immediately give in to the "international regulatory consensus" soon began to feel the power of their domestic multinationals. Despite strong resistance from industry and research, the German government continued, throughout the second half of the 1970s, to consider the introduction of special legislation for recombinant DNA work. When Hoechst, one of the world's largest chemical companies, announced large-scale research cooperation with the genetics department of Massachusetts General Hospital in 1980, the German government announced it now saw no more need for special legislation for recombinant DNA work.[44]

By the early 1980s, the public discussion of genetic engineering seemed to be "under control." An Organization for Economic Cooperation and Development (OECD) report comments on this period:

> In the early 1970s, when the technology of genetic manipulation was first acquired, predictions of scientists ranged from panacea to pandemic. A public storm followed and it was not at all surprising that national authorities in many countries, having been told that this technology was capable of creating new forms of life and that scientists themselves had requested a moratorium for this type of research, responded by setting up groups and committees to consider the social and political acceptability of the risk. A public feeling of instinctive mistrust towards scientists promoting genetic engineering was widespread. However, finally, after considerable public debate and advice from scientists, medical and epidemiological experts, the general conclusion has been reached that, provided suitable precautions are taken, the benefits of the technology far outweigh any conjectural risks.[45]

After forging a "hazard consensus" in the early 1980s, most OECD countries engaged in strategies to systematically promote biotechnology by means of a variety of policy instruments, from increased research funding for the universities to tax policies favoring investments in industry.[46] These policies reflected the situation in the world economy, putting pressures on states to adopt an offensive strategy in world markets, while the world system became more diffuse and decentralized. The policy discourse in the various industrialized countries constructed a sociopolitical scenario that made the rapid introduction and encouragement of biotechnology indispensable. Various external and internal developments that otherwise would constitute signs of disruption within the collective identity were made "foreign."[47] The epidemic dimension of cancerous diseases, environmental degradation, and the accumulation crisis were all removed from their complex sets of causes and became "evils," virtually justifying the deployment of new practices and new strategies of control as offered by the new biotechnology. The rewriting of life and individuality, the discourse of deficiency on the molecular level, and the rewriting of collective identity on the political level entered a mutually reinforcing relationship. Intervention, a strategy inherent in molecular biology since its origins in the 1930s, became its new defining modus operandi.

The U.S. leadership in the new biotechnology, vividly illustrated by booming venture capital driven "bio-boutiques," caused Japan and Western Europe in particular to face up to the biotechnology challenge. Whereas Japanese industrial policy shifted toward a stronger emphasis on knowledge-intensive technologies, European policymakers indulged in the rhetoric of an ever-widening technology gap between Europe, on the one hand, and the United States and Japan on the other. Despite great variations in strategies and rhetoric among the United States, Japan, and Europe, the early 1980s overall marked a dramatic acceleration of efforts in the various countries to encourage research and development in biotechnology, always with the intention to keep the particular country attractive as a research and industry location.

U.S. biotechnology policy focused on combining massive funding for "basic research" with a systematic policy of business encouragement. NIH support, particularly in the biological sciences, has been steady since the 1950s, and thus crucial to the strength of the American position. With the arrival of genetic engineering, this support has risen dramatically, making the U.S. government by far the largest sponsor of biomedical sciences worldwide. In 1982 and 1983, government funding of biotechnology was estimated at approximately $520 million per year.[48] By 1987, the estimated level had increased to $2.7 billion. And, according to the president's 1991 budget, roughly $5 billion was invested in biotechnology research, development, and manufacturing, of which approximately $3.2 billion was supported by the government. In 1992, then President George Bush selected biotechnology research for the fiscal year of 1993 as a budget initiative, with governmental expenditures planned to rise to $4 billion in 1993. The most significant agency to provide support for biotechnology came from the NIH, which in 1987 provided more than 87 percent of all government funding for biotechnology.[49] In the United States, a variety of measures encouraged what were termed "industry-university research partnerships." New federal patent legislation passed in 1980 relaxed criteria for federal approval of licensing agreements between universities and private businesses. The 1986 Technology Transfer Act prompted the NIH to establish guidelines giving companies exclusive licensing rights arising from federally funded research.[50]

Whereas the support for molecular biology and biotechnology in the United States has been steady and high, Japan and Europe practiced policies of catching up to what was perceived as a U.S. lead. The Japanese government's involvement in biotechnology goes back to 1971. However, it was not before 1980 that, in reaction to the publicity given to U.S. biotechnology companies and the value of their stocks, a series of government programs and initiatives was initiated.[51] The scope of the Japanese government's support was, however, minuscule compared to that of the United States: for 1983, the estimated total government support was around $50 million; for 1986, around $166 million; and, for 1989, $603 million.[52]

In Europe, a similar process can be observed. With the exception of Germany, whose biotechnology policy dates back to the early 1970s, most European governments set major biotechnology initiatives only in the early 1980s. France, for example, created a special section, the Mission Biotechnology, within the Ministry of Research. In 1982, a framework *programme mobilisateur* was set up. In Great Britain, a special Biotechnology Directorate was established within the Science and Engineering Council, and the Department of Trade and Industry as well as the University Grant Committee increased funding for biotechnology. On the level of the European Community, the policy focus rested on the removal of "divergent regulations" in the various membership countries, the creation of an effective patent system, and the development of a joint research-and-development program. However, a look at the total EC spending over the years reveals again the scale of the U.S. expenditures. The total EC expenditures in 1982 and 1983 are estimated at around $156 million, and for 1989 $366 million.[53] Clearly, the U.S. government outspent all other countries or trading blocs, while the

Japanese government rallied as the second-largest financier of biotechnology after the European Community member states.

But the technology race was soon supplemented by the emergence of a "more peaceful" system of transnational industrial relations in biotechnology. In fact, the image of nations engaged in a situation of competition coevolved with the articulation of a global economic space of biotechnology, thus imploding the meaning of a *national* biotechnology industry. The social construction of a "technology race" positioned private industry to take a decisive role in the politics of genetic engineering. In the United States, the scientific developments around genetic engineering were quickly exploited commercially. This process was located in start-up and venture-capital-driven companies, and involved significant resources from established manufacturing firms. In contrast, in Japan as well as in Europe, the biotechnology industry did not evolve around new start-up firms but around developed companies. However, both Japanese and then European companies were soon setting up collaborative links mainly with new U.S. start-up companies.[54] In general, in the biotechnology industry, as in other industries, the same crosscutting forces were encouraging national and international strategic alliances: high research costs and long development cycles; rapid technical change in the production sphere; the need to recoup increasing capital investments through wider and deeper market penetration; and a shifting international balance of economic power and technological prowess.[55] Hence, from very early on, alliances, particularly international alliances, have played a major role in the biotechnology strategies of large firms. Between 1981 and 1989, only seven out of 544 U.S. interfirm strategic alliances involved strategic alliances between two large companies.[56] This tendency toward alliances is a result of the structure of the U.S. biotechnology industry: of the approximately thirteen hundred U.S. biotechnology companies, 44 percent are very small firms, and another 36 percent are only slighter larger. Hence, for the majority of these firms, alliances are crucial for capitalizing their operations and spreading risks.[57] Japanese companies in particular went quickly into strategic alliances with these U.S. companies. Between 1980 and 1983 alone, Japanese companies entered into 188 biotech strategic alliances with U.S. firms. These alliances contributed to a rapid buildup in Japanese pharmaceutical and chemical companies.[58] Of the European countries, Germany, the United Kingdom, and Switzerland accounted for the 310 U.S. strategic overseas alliances (other than those with Japan) undertaken since 1981. This number reflects the strength of the chemical and pharmaceutical industries in these countries and, in general, in Europe. Again, it was mainly large European or large American firms that established alliance with smaller (U.S.)

partners. This reflected the strength of the U.S. start-up-based biotechnology industry and the gradual rise of the European biotechnology industry.[59]

With these processes of globalization in the biotechnology industry, the race between nations and trading blocs for industrial leadership turned increasingly into a race between nations for industrial investments and location. The new global/local economic space of biotechnology, opened up by the small firms in the early and mid-1970s, had become an important field for profit making and began to redefine the logic of national biotechnology policy making.[60] The decomposition of national economic spaces of biotechnology was paralleled by the articulation of biotechnology into the activity and responsibility of the territorial state.

However, at least in the past two decades, the global rush to exploit the economic promises of genetic engineering can hardly be construed as an impressive commercial success story. A number of factors are responsible for this gap between rhetoric and reality. Most important is the growing evidence of the limitations of some of the core assumptions made by the radically reductionist research tradition in molecular biology (which informs much of current biotechnology research and development). Reductionism in molecular biology represented biological systems in terms of the physical interactions of their parts. In this view, the fundamental understanding of biology comes only from the level of DNA, the alleged blueprint for living systems,[61] At the core of this scientific narrative is the following construction:

DNA → RNA → Protein → everything else, including disease

This scheme is also known as the "central dogma" of molecular biology: a sequence of DNA functions either by directly coding for a particular protein or by being necessary for the use of an adjacent segment that actually codes for the protein.[62] In the United States, for example, by the late 1980s the vast majority of the NIH research budget was going to projects that reflected this dogma.[63] Following the logic of a reductionist life narrative, many researchers in molecular biology truly believe that studies on gene cloning and regulation will provide important medical advances.

This genetic research paradigm, however, is questioned by a number of theoretical counterarguments, and also by findings of epidemiological studies. The goal of the genetic paradigm in medicine—to detect single-gene associations with human disease—finds itself challenged, for example, by the current patterns of disease distribution. For the major diseases like cancer, cardiovascular conditions, and most psychological illnesses, there is no evidence for either single-gene mutations or chromosomal aberrations as the unitary cause. Rather, the major diseases today are polygenetic and complex, have environmental determinants, and are not approachable by genetic analysis alone as suggested by the reductionist narrative of molecular biology.[64] The Human Genome Project argues that genetic effects may be separated from effects emerging from gene-gene, gene-gene product, and gene-environment interaction. Furthermore, the assumption is made that genetic programs are operative and that polygenic diseases can be mapped onto Mendelian components, or onto grouped components. But most of these assumptions have been rejected by population geneticists and are gradually being rejected by biologists working on fundamental problems of molecular and cell biology.[65]

One of the main criticisms leveled against the Human Genome Project is that its framing of disease ignores the complexity of the issues and the limitations that it can impose on the possibility of such theoretical computation. A crucial example of such difficulties is the central protein-folding problem. In the representation of molecular biology, the assumption is that the primary structure or amino acid sequence of a protein determines its tertiary structure or three-dimensional conformation. Therefore, it should be possible to calculate the conformation from the sequence information. However, despite three decades of research efforts, the protein-folding problem has yet to be solved.[66] But protein conformation is the key to receptor studies that are essential for successful drug design.

Such "research bottlenecks," biological counternarratives, and patterns of epidemology have become part of a critique that not only questions biotechnology as a high technology, but, on a more fundamental level, raises doubts about its theoretical foundations and related predictions on the transformative potentials of genetic engineering. On the level of product development, this constellation has given rise to talk of biotechnology's "long and winding road to the marketplace."[67] Failures and the limited successes in product development in the United States constituted important destabilizations of biotechnology policy worldwide. After the period of gross exaggerations by scientists, venture capitalists, and entrepreneurs in the late 1970s, in the 1990s a more sober mood set in, which was partially caused by the disappointing achievements of the biotechnology industry.[68]

Globalization, Democracy, Genetic Engineering

So, who is "us"? After a look at the shifting diagram of social forces on the level of production, state, and world order, and their relationship with biological discourse, we understand Robert B. Reich better when he tells us that "we" are not "the corporations" but "the producers." As the politics of transnational companies supersedes the politics of the nation-state, liberal thinkers are challenged to rethink the contours of democracy.

Since the 1970s, the hegemonic world order experienced a process of disintegration. With the demise of the centrality of the United States, a more decentralized pattern of the international world order emerged, with Europe and Japan increasing their economic weight and transnational corporations assuming unprecedented influence. The ability of states to control their economies has decreased dramatically. Trade unions and labor parties have been forced into defensive strategies as unemployment rose. While the United States faces the painful task of confronting the social impact of the Reagan-Bush years, in Europe the rhetoric of "Euro-sclerosis," with the attendant deregulatory recommendations, is on the rise again.[69]

The construction of the new economic space of biotechnology must be seen in this context. With the emergence of molecular biology, a new mode of representing life had been shaped. Ultimately, this system of representation provided the interventionist means to construct information-bearing molecules that no longer need to preexist within the organism.[70] The inscription of molecular biology's frameworks of signification into policy translated into biotechnology policy with the "discourse of deficiency" at its core. Just as in the Human Genome Project's quest for health translated into a search for the genetic basis of unhealth, "normality" came to be defined by the absence of those alleles said to cause disease.[71] Other expressions of life, such as plants and animals, became "extracellular projects" and were rewritten in terms of "absences," of "improvables" needing the intervention of genetic technologies. This rewriting of life found itself mainly expressed in the fields of basic research and culture, such as in increasingly dominant images of disease causation. Nevertheless, in this process research laboratories, field-test areas, hospitals, and farms became the terrain for practices geared toward interventions on the subcellular level, a process that reflected the rise of molecular biology as a culturally hegemonic project. Although the grandiose biotechnology scenarios devised over the past two decades apparently failed to produce the anticipated outcomes, they nevertheless did something else, and that something else displayed its own logic in its coproduction of the social world.

Social resistance against molecular biology's writing/rewriting of life/subjectivity is inextricably related to the rise of molecular biology to its dominant status in the life sciences. In the absence of more than a handful of products of the biotechnology industry with a mass market, much of the critique of genetic engineering focused on potential scenarios of the impact of genetic technologies and issues of culture and reality definition. The interrelated patterns of change in scientific practice, production, subjectivity, governmental policies, and systems of surveillance have become fields of contestation and intervention by social movements. This resistance is closely related to a process of political mobilization that emerged in the 1960s and to the rise of social movements challenging the political order.

Particularly in the past few decades, many countries experienced a vigorous growth of new social movements. As mentioned earlier, the increased internationalization of production had considerably reduced the capabilities of nation-states to autonomously steer and control the economy and society. Several factors rendered the distinction between state and civil society even more unclear: the conflicts and contradictions of advanced industrial society increasingly created phenomena of "ungovernability"; neoconservatism was on the rise; and the post-World War II model of economic growth and social compromise came under attack. In the past, democracy was defined by the separation of the state from civil society, a system where the state translated the private interests formed in civil society into terms of public institutions. With public policies having increasingly significant impact on citizens, the core institutions of civil society, such as political parties, have been disintegrating, and the autonomous capacities of the state are decreasing. In the past decade, clear signs of a reconstitution of civil society by new social movements have become very visible. Their intervention has constituted a reconsideration of the patterns of a disintegrating domestic as well as international hegemony of the political order and, at the same time, manifested a probing of the "politics of individuality" underlying the hegemonic or dominant order.

By radically questioning definitions of codes and nominations of reality, social movements have challenged the dominant logic on symbolic grounds. Rules of normality, spaces of difference, and frames of reality have been thematized by a variety of movements, from the women's movement to the ecological movement. Their agendas are not necessarily rooted in their access to power, but in challenging the logic of the dominant rationalizing apparatuses.[72] At the same time, this politics of meaning is closely tied with recent social conflicts related to new modes of exploitation in the emerging neoliberal world order and attendant strategies of surveillance. Genetic engineering, with its broad scope of potential interventions, is today inextricably interconnected to a multitude of resistances. They focus on the few available products of the biotechnology industry, such as genetic screening tests in the workplace and the use of genetic engineering methods in milk production, in particular the application of rBST (recombinant bovine somatropin, a hormone that increases the cow's milk production). But resistance can also be found on the more hypothetical/symbolic level, as in the rejection of genetic screening and therapy by disabled groups who fear that these technologies will make them into "others." Images of Nazi eugenics, Third World exploitation, and rBST-tortured cows invaded the biotechnology discourse, creating counterdiscursive ruptures, blurrings, and mixtures. In this reading, "the truth" of molecular biology was rewritten as an ideology; genetic engineering was contextualized within the history of eugenics; and agricultural biotechnology was placed within the framework of the exploitation of the non-Western world. Molecular biology's system of signification came to be associated with a strategy of cultural infiltration that changed and normalized the parameters of human self-recognition and understanding of nature, a strategy countered by a politics that allowed differences to establish space for themselves as alter identities.

In the United States, groups such as Friends of the Earth, the Committee for Responsible Genetics, and Jeremy Rifkin's Foundation on Economic Trends engaged early on in a discourse of resistance. Similar developments happened in Japan with groups such as women and disabled against eugenic policy, consumer movements concerned with food safety, and Citizens Concerned about Genetic Engineering that questioned trends in biotechnology policy and regulation. In Europe, starting in Denmark and Germany, a broad mobilization against genetic engineering began in the mid-1980s. Similarly, a multitude of non-Western movements, such as India's Gene Campaign and Zimbabwe's environmental network ZERO, began to mobilize internationally against biotechnology. These non-Western groups were partially supported by and worked in cooperation with Western nongovernmental organizations, such as GRAIN (Genetic Resources Action International) and the German Gen-Ethic Network.[73]

The cultural struggle of the many movements operating worldwide took the political form of confrontations over particular pieces of legislation, over the distribution of financial resources in research policies, or over engagement in the development of alternative forms of land cultivation. In Europe particularly, the strength of the new social movements and Green parties was to a considerable extent responsible for the adoption of comparatively stringent regulations on contained use and deliberate release of genetically modified organisms.[74] Groups in the United States, such as the Committee for Responsible Genetics and the Foundation on Economic Trends, have persistently sought greater public involvement in the political decision-making process concerning genetic engineering but have had a more limited impact than similar groups in Europe.

Social resistance against genetic engineering was met by counterstrategies seeking to establish a framing of biotechnology as an articulation of progress and modernization. As the quest for limits to the freedom of research and development rose after the Asilomar conference on recombinant DNA research in the mid-1970s, the iconography of technology races, economic viability, and medical progress represented the question of the social, political, and ecological desirability of genetic engineering as a nonnegotiable necessity. These struggles over the meaning of biotechnology find themselves inscribed in findings of public-opinion survey research. Cross-national research on public opinion on biotechnology displays a public seriously concerned about the potential risks of biotechnology and, at the same time, appreciating its potential benefits.[75] These polls reflect the dominant representations of biotechnology as a technology to which there is no alternative. This view allows for an unresolved cognitive stabilization of biotechnology as at once risky and beneficial, a crucial prerequisite for the hegemonic status of molecular biology. Not surprisingly, political strategies against the outspoken critics of genetic engineering are usually based on an exploitation of this ambiguity in public opinion, vilifying opposition and critique as "irrational" and interpellating the public as the hegemonic field where concern and appraisal of biotechnology can peacefully coexist without substantially inhibiting the further expansion of genetic technologization.

The impact of social resistance against genetic engineering can be seen worldwide and can be found in different articulations, depending on political constellations. In Europe, the EC-wide ban on rBST, the delay of the European Human Genome Initiative, and the blocking of European patent legislation by the European Parliament, as well as the regulation of contained use and deliberate release of genetically modified organisms, are some of the successes attributable to a considerable degree to the influence of international networking movements and Green parties. In the United States, for example, the Food and Drug Administration (FDA) approved rBST in 1994; however, in the wake of consumer protest, Vermont, Wisconsin, and Maine passed laws requiring the labeling of milk from rBST-treated cows, and grocery stores around the country sprouted signs over their milk racks reading "rBST-free milk." As the trade journal Bio/Technology put it: "the BST mess may well cause a backlash against genetically engineered foods in general and frighten an increasingly indecisive FDA into clamping down further in its regulatory reviews of recombinant foods and agricultural products."[76] In the Third World, a multitude of local and regional activities exploring new models of sustainable development demonstrate increasing resistance to trends in agricultural biotechnology, such as the arrival at the marketplace of genetically engineered hybrid seed packages, often requiring expensive chemical pest controls and fertilizers to ensure optimum yield. In the Philippines, to give one example, a community group, together with scientists from the University at Los Baños, launched a plant-breeding program called MASOPAG. The project seeks to recover and preserve the varieties of rice on lands that have lost much of their genetic diversity since the early 1970s, during which time the so-called high-yielding varieties that require fertilizer and chemical pesticides have gained ground. MASOPAG is striving to reverse this trend and build on the farmer innovation that has traditionally conserved and increased the diversity of crop varieties.[77] Today, such initiatives to protect local knowledge against the competition from globally operating high-tech agricultural biotechnology companies are widespread in many Third World countries. On the level of international politics, these initiatives gained visibility at the 1992 alternative Rio Earth Summit, where an "alternative" BioConvention was passed deploring the omission of germplasm collections from the Convention on Biological Diversity signed at the Rio Summit.[78]

This points to an important development in the contested shaping of genetic technologies. If we can read new social movements as an effort to reconstitute civil society, the reconstitution takes the form of a repoliticization of the institutions of civil society, such as work, production, and science. This repoliticization of civil society assumes increasingly global dimensions. The new global political spaces emerged along networks of economic, social, and cultural relations in situations of activation for particular political purposes.[79] The symbolic challenge of genetic engineering must be seen as part of this emerging global project.

As the identification with the nation-state increasingly erodes, alternative constructions of collective identity gain in importance. Certainly, liberalism in its many variations offers various alternative solutions, such as consumerism or, in Reich's case, productivism. But it is exactly this political imaginary and its underlying policies of subjectivity that are critically explored and deconstructed in the emerging global civil society. As discrepancies arise between socially available identities and those "overflowing" on the individual human level, the construction of spaces for the reflection on alter identities introduces a new dynamic into the interrelationship between genetic engineering, democracy, and identity.

Notes

1. I am grateful for helpful comments to Lily E. Kay, Peter Taylor, Chris Cuomo, and Randy Martin.
2. Robert B. Reich, "Who Is Us?" *Harvard Business Review* (January–February 1990): 53–64, 53.
3. Ibid., 54.
4. William E. Connolly, *Identity/Difference: Democratic Negotiations of Political Paradox* (Ithaca, N.Y.: Cornell University Press, 1991), 64.
5. John Hodgson, "An Appetite for Technology: Hoffmann-La Roche," *Bio/Technology* 10 (1992): 86–88.
6. Commission of the European Communities, *Proposal for a Council Decision Adopting a Specific Research Programme in the Field of Health: Predictive Medicine: Human Genome Analysis (1989–1991)*, COM(88) 424 final SYN 146 (Brussels, July 20, 1988), 3.
7. For an excellent exposition of contemporary panopticism, see Stephen Gill, "The Global Panopticon," manuscript, Department of Political Science, York University, Toronto, 1993.
8. Barry Smart, "The Politics of Truth," in *Foucault: A Critical Reader*, ed. David Couzens Hoy (Cambridge, Mass.: Basil Blackwell, 1992), 157–73, 161.
9. Colin Gordon, "Governmental Rationality: An Introduction," *The Foucault Effect: Studies in Governmentality with Two Lectures by and an Interview with Michel Foucault*, in Graham Burchell, Colin Gordon, and Peter Miller (Chicago: University of Chicago Press, 1991), 1–51, 20.
10. Michel Foucault, *The History of Sexuality, vol. 1, An Introduction*, trans. Robert Hurley (New York: Vintage Books, 1980), 139.
11. Lily E. Kay, *The Molecular Vision of Life: Caltech, the Rockefeller Foundation, and the Rise of the New Biology* (New York: Oxford University Press, 1993), 49.
12. Ibid., 9.
13. Stephen Gill, "Reflections on Global Order and Sociohistorical Time," *Alternatives* 16 (1991): 275–314, 282.
14. Thomas Ferguson, "Industrial Conflict and the Coming of the New Deal: The Triumph of Multinational Liberalism in America," in *The Rise and Fall of the New Deal Order, 1930–1980*, ed. Steve Fraser and Gary Gerstle (Princeton, N.J.: Princeton University Press, 1989), 3–31, 7.
15. Kay, *The Molecular Vision of Life*, 26.
16. Smart, *The Politics of Truth*, 162.
17. Gill, "Reflections on Global Order," 283.
18. Robert W. Cox, *Production, Power, and World Order: Social Forces in the Making of History* (New York: Columbia University Press, 1987), 212.
19. Gordon, "Governmental Rationality," 20; David Campbell, *Writing Security: United States Foreign Policy and the Politics of Identity* (Minneapolis: University of Minnesota Press, 1992).
20. Charles F. Andrain, *Social Policies in Western Industrial Societies* (Berkeley: Institute of International Studies, 1985), 45.
21. Rod Coombs, Paolo Saviotti, and Vivien Walsh, *Economics and Technological Change* (Totowa, N.J.: Rowman and Littlefield, 1987), 223–29.
22. For overviews of the evolution of science and technology policy, see J. Ronayne, *Science in Government* (Baltimore: Arnold, 1984); Ros Herman, *The European Scientific Community* (Harlow, Essex: Longman, 1986).

23. Philip J. Pauly, "Essay Review: The Eugenics Industry—Growth or Restructuring?" *Journal of the History of Biology* 26 (1993): 131–45, 145.
24. Edward Yoxen, "Life as a Productive Force: Capitalising the Science and Technology of Molecular Biology," in *Studies in the Labour Process*, ed. Robert M. Young and Les Levidow, vol.1 (London: CSE Books, 1981), 95.
25. Ibid., 98.
26. Interview, by the author, Volkswagensstiftung, Hannover, 1991.
27. Yoxen, "Life as a Productive Force," 98–101; Pnina Abir-Am, "From Multidisciplinary Collaboration to Transnational Objectivity: International Space as Constitutive of Molecular Biology, 1930-1970," in *Denationalizing Science: The Contexts of International Scientific Practice*, ed. Elisabeth Crawford, Terry Shinn, and Svesker Sorlin (Dordrecht: Kluwer Academic Publishers, 1993), 153–86, 172–76.
28. Abir-Am, "From Multidisciplinary Collaboration to Transnational Objectivity," 153–86.
29. Gordon Wolstenholme, ed., *Man and His Future* (Boston: Little, Brown, 1963).
30. Hans-Jörg Rheinberger, "Genetic Engineering and the Practice of Molecular Biology" (paper given at the Mellon workshop on the topic "Genetic Engineering: Transformations in Science, Politics and Culture," MIT, April 30 and May 1, 1993), 8.
31. Gill, "Reflections on Global Order," 284–89.
32. Patrick Camiller, "Beyond 1992: The Left and Europe," *New Left Review* 175 (1989): 5–17, 7.
33. Gill, "Reflections on Global Order," 289.
34. Cox, *Production, Power, and World Order*, 281–86.
35. Commission of the European Communities, European File, *Tomorrow's Bio-Society* (Brussels, 1980), 2.
36. Diane J. McLaren, "Human Genome Research: A Review of European and International Contributions" (unpublished manuscript, London: Medical Research Council, 1991).
37. Leroy Hood, "Biology and Medicine in the Twenty-First Century," in *The Code of Codes: Scientific and Social Issues in the Human Genome Project*, ed. Daniel J. Kevles and Leroy Hood (Cambridge: Harvard University Press, 1992), 136–63, 137–38.
38. James Der Derian, *Antidiplomacy: Spies, Terror, Speed, and War* (Cambridge, Mass.: Blackwell Publishers, 1992), 32.
39. Gill, "The Global Panopticon." See also David Lyon, *The Electronic Eye: The Rise of Surveillance Society* (Minneapolis: University of Minnesota Press, 1994).
40. See Dorothy Nelkin, "The Social Power of Genetic Information," in *The Code of Codes*, ed. Kevles and Hood, 177–90.
41. Dorothy Nelkin and Laurence Tancredi, *Dangerous Diagnostics: The Social Power of Biological Information* (New York: Basic Books, 1989), 169.
42. DG XII/Commission of the European Communities, *FAST Subprogamme C "Bio-Society,"* FAST-ACPM/79/14-3E (Brussels, 1979).
43. John Tooze, "Emerging Attitudes and Policies in Europe" (memo, Libraries of MIT, Institute Archives and Special Collections, Recombinant DNA Collection, n.d.).
44. Heinz Theisen, "Zur Demokratieverträglichkeit der Bio- und Gentechnologie, *Soziale Welt* 42 (1991): 109–30, 113.
45. Alan Bull, Geoffrey Holt, and Malcolm D. Lilly, *Biotechnology: International Trends and Perspectives* (Paris: OECD, 1982), 60.
46. Organization for Economic Cooperation and Development, *Biotechnology and the Changing Role of Government* (Paris: OECD, 1988).
47. For a general discussion of this political strategy, see Connolly, *Identity/Difference*, 207.
48. This figure as well as all following figures concerning governmental expenditures should be read with great caution. Different research and government agencies, re-

ports, and so on have different definitions of what counts as biotechnology research. This limits the comparability of the data issued. However, the data at least reflect the grand lines of the expenditure structures.

49. Suzanne S. Groet, "Biotechnology and the U.S. Government: The Pot at the End of the Rainbow?" in *The Business of Biotechnology: From Bench to the Street*, ed. R. Dana Ono (Boston: Butterworth/Heinemann, 1991), 199–211, 203; Federal Council for Science, Engineering, and Technology/Committee on Life Sciences and Health, *Biotechnology for the 21st Century* (Washington, D.C.: Federal Council for Science, Engineering, and Technology/Committee on Life Sciences and Health, 1992), 18.
50. Sheldon Krimsky, *Biotechnics and Society: The Rise of Industrial Genetics* (New York: Praeger Publishers, 1991), 66–68.
51. Robert T. Yuan and Mark Dibner, *Japanese Biotechnology: A Comprehensive Study of Government Policy, R & D and Industry* (Basingstoke: Macmillan, 1990), 19–20.
52. European Community, Directorate-General XII, "National Initiatives for the Support of Biotechnology," (Brussels, n.d.), 8, CUBE-Documentation, *National Biotechnology Policy—Japan* (Brussels, 1988), 1; Yuan and Dibner, *Japanese Biotechnology*, 24.
53. European Community, "National Initiatives for the Support of Biotechnology," 34; CUBE, "National Biotech Policy: EC Member State Review," (Brussels, 1992), 1.
54. V. Walsh, I. Galimberti, J. Gill, A. Richards, and Y. Sharma, *The Globalisation of Technology and the Economy: Implications and Consequences for the Scientific and Technology Policy of the E.C.*, Monitor—FAST Programme, Prospective Dossier No. 2, "Globalisation of Economy and Technology" (Brussels: n.p., 1991), 19–20.
55. Sandor L. Boysen, "Market Revolution: The Explosion of Biotechnology Strategic Alliances" (prepared for the United Nations Center for Science and Technology for Development, September 14, 1989), 17.
56. Ibid., 29.
57. Ibid., 31.
58. Ibid., 34.
59. Ibid., 36–41.
60. Martin Kenny, "The Creation of a New Economic Space: The Commercialization of Molecular Biology" (paper given at the Chemical Heritage Foundation Conference "Private Science: The Biotechnology Industry and the Rise of Contemporary Molecular Biology," Philadelphia, October 28–30, 1993).
61. Alfred I. Tauber and Sahota Sarkar, "The Human Genome Project: Has Blind Reductionism Gone Too Far?" *Perspectives in Biology and Medicine* 35 (1992): 222–35, 228.
62. According to the "dogma," three types of processes are responsible for the inheritance of genetic information and for its conversion from one form to another: (1) information is perpetuated by replication; a double-stranded nucleic acid is duplicated to give identical copies; information is expressed by a two-stage process; (2) transcription generates a single-stranded RNA identical in sequence with one of the strands of the duplex DNA; (3) translation converts the nucleotide sequence of the RNA into the sequence of amino acids comprising a protein. See Benjamin Lewin, *Genes IV* (Oxford: Oxford University Press, 1990), 109–13.
63. Richard C. Strohman, "Ancient Genomes, Wise Bodies, Unhealthy People: Limits of a Genetic Paradigm in Biology and Medicine," *Perspectives in Biology and Medicine* 37 (1993): 112–45, 117.
64. Ibid., 119–20.
65. Ibid., 130.
66. Tauber and Sarkar, "The Human Genome Project," 223.
67. Ernst and Young, *Biotech '95. Reform, Restructure, Renewal. Industry Annual Report* (1994), n.p.

68. See Robert Teitelman, *Profits of Science: The American Marriage of Business and Technology* (New York: Basic Books, 1994), 192–95.
69. See, for example, "Community's Social Action Plans Succumb to Sabotage and Recession," *Financial Times*, November 19, 1992; "Labor Market Gripped by Euro-Sclerosis," *Financial Times*, June 21, 1993.
70. Rheinberger, "Genetic Engineering and the Practice of Molecular Biology," 7.
71. Evelyn Fox Keller, "Nature, Nurture, and the Human Genome Project," in *The Code of Codes*, ed. Kevles and Hood, 281–99, 298.
72. Alberto Melucci, "The Symbolic Challenge of Contemporary Movements," *Social Research* 52 (1985): 789–816, 810.
73. Henk Hobbelink, "Für die Menschen in der Dritten Welt kommt die Biotechnologie als integriertes Paket," *Gen-Ethischer Informationsdienst* 6 (1990): 11–14; Ute Sprenger, "Ob Nord oder Süd-Vielfalt reduziert die Risken in der Landwirtschaft. Praktischen Inititativen gegen die genetische Monotonie," *Gen-Ethischer Informationsdienst* 6 (1992): 13–17.
74. For Germany, see Herbert Gottweis, "German Politics of Genetic Engineering and Its Deconstruction," *Social Studies of Science* 25 (1995), 195–235.
75. Berhard Zechendorf, "Public Opinion on Biotechnology: A Comparative Analysis" (Brussels: European Commission, 1991).
76. Russ Hoyle, "FDA's BST Policy Wreaks Havoc in the Market," *Bio/Technology* 12 (1994): 570–71, 570.
77. René Salazar, "MASOPAG: Alternative Community Rice-Breeding in the Philippines," *Appropriate Technology* 18 (1992): 20–21.
78. Diversity 8 (1992): 3.
79. Ronnie D. Lipschutz, "Reconstructing World Politics: The Emergence of Global Civil Society," *Millennium: Journal of International Studies* 21 (1992): 389–420, 393.

Contradictions along the Commodity Road to Environmental Stabilization: Foresting Gambian Gardens

Richard A. Schroeder

One person's degradation is another's accumulation.
▸ *Piers Blaikie and Harold Brookfield (1987)*[1]

Foresters, soil scientists, wildlife preservationists, and other would-be reclamation specialists, whose agenda it is to "reverse" processes of environmental degradation and return "nature" to a state of relative equilibrium, are confronted with a paradox.[2] It is often not enough to simply stop or avoid practices leading to degradation; restoring and preserving the integrity of "nature" involves subjecting it to human control.[3] The trouble is that remedial efforts that involve human intervention such as reforestation and terracing tend to require significant investments of labor and are quite costly, while at best affording only deferred economic benefits.[4] This expense inhibits the individual "land manager" from assuming responsibility for remediation on his or her own account, and forces the state, by default, to carry out the socially necessary tasks of environmental production, protection, and restoration. Under such circumstances, questions of access to, and control of, resources come into play, and challenges to the legitimacy of state action are frequent: "There is a constant likelihood that [state-imposed stabilization practices] will be seen . . . as coercion."[5] This prospect leaves environmentalists in a quandary. When critical resources are threatened, the use of forceful means to restrict specific practices seems justified, yet heavy-handed approaches that privilege some users (e.g., ecotourists) over others often prove counterproductive: "Undue inequalities in access to resources lead to poaching and expensive protection, and simply diverts [*sic*] pressure on to resources elsewhere."[6]

Not surprisingly, then, state managers and environment-oriented developers have stepped up efforts in recent years to chart a different course. In contrast to both regulatory efforts, which engender opposition by criminalizing traditional land-use practices and penalizing resource users who engage in them, and more coercive measures, which subject resource managers to police and (para)military force,[7] environmentalists also rely heavily on systems of *positive* economic incentives to promote sound land-use practices; they frequently opt for what might be called the "commodity road" to stabilization.

According to this approach, the marketing of commodities harvested or otherwise gleaned during the initial or intermediate stages of slow-developing stabilization efforts provides short-term payoffs that enhance both the total and the rate of return on investment in those systems. Local involvement is thus encouraged, and many of the political problems of externally imposed solutions are theoretically minimized. Policies centered on "extractive reserves," for example, and parallel efforts at "green marketing," involve negotiating limited access to, and the controlled removal of, key commodities from old-growth forests and other ecologically sensitive areas by indigenous groups.[8] The regulation of such extraction purportedly converts land to productive uses, thus satisfying (or at least holding at bay) interests that advocate a more invasive and wholesale appropriation of assets on economic grounds.[9] At the same time, the protection of local livelihood strategies preserves the detailed environmental knowledge of indigenous groups, a cultural legacy that is deemed crucial for long-term survival of reserve area ecologies.[10]

Although such policies may achieve a certain measure of success regarding the interim goal of income generation, any sense that the commodity road inevitably results in a more stable resource base must be quickly dispelled. As Roderick P. Neumann demonstrates, the investment opportunities created in the context of privatized environmental initiatives such as ecotourism often displace local users and give rise to patterns of accumulation that have little to do with resource stabilization per se.[11] The critical question posed by commodity road strategies, then, is whether, in their efforts to accelerate natural rejuvenation by promoting commodity production, environmental developers do not unleash social processes that are themselves in some way counterproductive to environmental goals.

The empirical evidence I use to explore these ideas is drawn from ethnographic research on agroforestry systems in The Gambia's North Bank Division in 1989 and 1991. This research emphasizes contradictions that have arisen in the implementation of two ostensibly progressive rural development policies, the first directed at gender equity and the second emphasizing environmental stability. In particular, I emphasize the tendency toward a reduction of species diversity in many commodity-based systems,[12] and a parallel trend toward more exclusive landholding rights. I argue that these social and ecological contradictions are not simply contingent outcomes that derive from the particulars of the Gambian case, but structural tendencies that grow directly out of the profit-taking ethos of commodity production.[13]

The first section below traces the outlines of a boom in market gardening that has left women in rural Gambia in a position to earn cash incomes that in many cases outstrip those of their husbands. The second describes a series of concurrent developments in agroforestry programs initiated by the Gambian state and multilateral donors. The adoption of a commodity-based stabilization strategy and the ensuing rise of a group of tree crop entrepreneurs have threatened existing horticultural enclaves as gardens are eclipsed by orchard projects. The third section examines closely the social relations that have spun out of the garden/orchard conflicts. Land and tree tenure provisions, and the specific interests and motivations of gardeners, landholders, and developers, are considered in turn in order to assess the net effect of the political-ecological interventions made in the region to date. The concluding section then returns to the broader questions concerning the efficacy of commodity road strategies for the creation of an environmental policy that is both ecologically stabilizing and politically empowering.

Gender Equity and The Gambia's Market Garden Boom

In the two decades since its formal adoption at a UN conference in Mexico City in 1975, the feminist-inspired "Women in Development" (WID) agenda has achieved considerable prominence in Gambian national politics. Aided in no small measure by the international embarrassment suffered by the Gambian government in the aftermath of failed gender-sensitive irrigation projects in the early 1980s,[14] advocates of more equitable development planning have succeeded in placing explicitly gender-sensitive goals at the center of several key planning initiatives.[15] Among the most notable is a 1991 World Bank project established at the personal behest of the Gambian president. This effort, a five-year, $15 million multisectoral program, was, at the time of its establishment, the only "freestanding" WID program funded by the bank anywhere in the world.

One of the centerpieces of the World Bank and other WID-oriented strategies has been women's horticulture. Over the past two decades, hundreds of women's market gardens have been established along the Gambia River Basin in an attempt to redress serious deficiencies in maternal and child welfare.[16] These initiatives are typically funded through small WID-oriented grants and loans, which are used to construct new garden infrastructure (wells and fences) and purchase tools and seed.[17] Their immediate objectives have been both income generation by women and nutritional benefits for their families. While the dietary impact of market gardens is somewhat unclear, thousands of families now depend on the sale of fresh cabbage, onions, tomatoes, peppers, and a wide variety of other vegetables as a primary source of cash income.[18] The economic effects of the garden boom have been especially significant in light of the declining fortunes of rain-fed cash crops in recent years. Groundnut (peanut) production, the principal source of cash crop income for male farmers in the region, has suffered badly due to the combined effects of two decades of poor rainfall, the removal of producer price and fertilizer subsidies as a condition of national debt rescheduling, and extremely poor terms of trade on international markets for oilseed products.[19] Consequently, the task of household reproduction now hinges on vegetable sales in many areas. As one male farmer from a North Bank garden district confided: "Gardening generates a greater benefit than the peanut crop that we [men] cultivate. Before you offer any help to people farming groundnuts, it is better you help [women] doing gardening, because we are using gardening to survive."

The Rise of the Tree Crop Entrepreneur

Support for the market garden boom was not the only major rural development initiative that grew out of the economic circumstances of the 1970s and 1980s. While women's horticultural perimeters were being constructed to help rural families weather a poor financial climate, several environmental initiatives were also undertaken by the Gambian state and international donors. Researchers have established that a gradual deforestation of the Gambian River Basin has been under way since approximately the 1940s. One study, for example, indicated that the extent of "closed woodland" in The Gambia dropped from 61 percent to 9 percent from 1946 to 1980.[20] Accordingly, in 1977, the Banjul Declaration was issued, calling for conservation strategies that would protect "for now and posterity as wide a spectrum as possible of our remaining fauna and flora."[21] In the same year, the Forest Act and the Wildlife Conservation Acts were signed, and in 1978 the Wildlife Regulations were enacted. Among the policies that soon followed were the reassignment of control over forest lands to the newly created Forestry Department, forestry extension programs organized through schoolteachers and the Department of Agriculture, the first annual national tree-planting campaign, and the banning of charcoal production.[22]

A particularly ambitious and well-funded strategy, begun under the auspices of USAID (the USAID Forestry Project, 1979–86), gave evidence of a concerted effort by forestry planners to promote environmental objectives through the use of commercial incentives. This effort encouraged Gambian villagers to replace fuel gleaned from dwindling "natural" forest reserves with wood produced on small communal woodlots located on the immediate outskirts of villages. In addition to subsidies for wells and fences, commodity sales incentives were also used by promoters who urged woodlot growers to sell surplus wood and forest products (e.g., fodder) for cash income. In this instance, the commodity road strategy failed to produce a significant shift in village-level land-use practices, ultimately, analysts concurred, because "the wrong tree species had been selected."[23] These species, primarily *Gmelina arborea*, neem (*Asadirachta indica*), and eucalyptus, were "wrong" because alternative fuel-wood species preferred by Gambians (*Pterocarpus erinaceus*, *Terminalia macropthera*, *Hannoa undulata*, *Combretum spp.*, *Prosopis africana*, and mangroves) were still available. Moreover, the effort to build a stabilization strategy around forest species ran up against what was termed a "time factor":

> Compared to most agricultural operations, tree growing is a long-term exercise. First benefits from fruit trees (Mango, Cashew, Citrus) can be expected in three to five years after planting; from fast growing exotic species (*Gmelina*, Neem, Eucalyptus) in five to ten years; and from indigenous species in ten to fifteen years. Some projects in The Gambia (orchard/woodlot projects, horticultural projects, beekeeping) indicate that as long as villagers are able to obtain intermediate benefits such as vegetable[s], fruits, honey or forage they are more prepared to accept the delay in harvesting [of timber and other forest products].[24]

In effect, analysts argued that the woodlot program failed because it simply did not carry the commodity road rationale far enough. Exotic species could be, and were to some extent, commodified, but the economic returns they brought were still too slow to warrant investment of time and capital by would-be orchard owners.

Rights to project benefits were also unclear. Lawry deems such ambiguity a "classic" problem with woodlot projects: "Projects usually provide that benefits will be distributed among the villagers, but the formula for distribution is not thought through, or is left to be determined . . . by 'the village.'" Such proposals ignore the fact that "'the village' is not the harmonious social formation that many project designers assume it to be. . . . To say that the project will benefit 'the village' overlooks the fact that villages are made up of individuals, families and groups with different goals and expectations, which are not necessarily equitably served by any given project."[25] The Gambian woodlot program was also based on the erroneous assumption that villagers would willingly contribute labor despite the fact that the issue of benefits distribution was left unresolved. Lawry provides a case in point:

> There the village *alkalo* [head], a very entrepreneurial farmer and fruit grower, made available land for the woodlot, but was unable to mobilize villagers to participate in planting, cultivation, and harvesting tasks. While [he] attributed this to lack of foresight by villagers, it was clear that [he] was retaining tight control over woodlot management. . . . villagers were uncertain over their rights . . . in relation to their labor contributions.[26]

The characterization of villagers in this instance as lacking in "foresight" is telling. By demonizing villagers, landholders simultaneously pay lip service to the prevailing developmental ideology and reproduce their privileged positions as arbiters of the distribution of developmental largesse. The fact that this tack suits their narrower entrepreneurial objectives is transparent.

When the next major national forestry initiative was mounted by the European Economic Community in 1986, program emphases shifted toward fruit production and resolution of the labor question, both in order to expedite tree planting. By this time, the garden boom was well under way, and low-lying garden sites had become attractive locational options for foresters because they offered secure fence enclosures, improved soils, and ready access to water. Crucially, they also contained a captive, if not wholly cooperative, labor force to water trees, manure plots, repair fences, guard against livestock incursions, and maintain wells. By the late 1980s, therefore, a firm consensus had formed among the Forestry Department, nongovernmental organizations (NGOs), and donors around the triple foci of (1) using the production of fruit commodities as the vehicle to promote reforestation, (2) concentrating tree-planting projects within gardens, and (3) managing the whole endeavor on the premise of being able to exploit a female labor reserve.

Although the resulting mix of property and labor claims was not entirely without precedent in The Gambia, the heavy emphasis on *female* labor to carry out tree-planting objectives was striking.[27] A United Nations Development Program (UNDP) official gave voice to the ideological basis for pinning the hopes of Gambian environmental stabilization efforts on women: "Women are the sole conservators of the land. . . . the willingness of women to participate in natural resource management is greater than that of men. Women are always willing to work in groups and these groups can be formed for conservation purposes."[28] The official's thoughts were echoed in promotional literature circulated by an NGO active in tree planting on the North Bank in 1991: "The involvement of women in the development process is vital in the Gambian context if sustainable development is the ultimate goal. . . . In The Gambia, [the Worldview International Foundation's] primary focus has been on women. . . . In the implementation of an environmental programme in the country, *they could be deemed the most precious and vital local resource*."[29]

Planning documents for a proposed $18 million USAID-sponsored Agricultural and Natural Resource (ANR) program also commented on the important role played by women: "The role of women in improved ANR is critical as women frequently have access to [read: are limited to] more marginal/degraded lands than men, and have exhibited a great deal of interest in conservation techniques. . . . Targeted conservation committees will continue to require a percentage of female members to assure women's participation."[30]

Although these references would seem to bear the marks of a critical environmental feminist ideology, the specific *locational* decision to target women's gardens has a more instrumental political-economic rationale.[31] One developer summarized the prevailing sentiment: "Women are reportedly *not good at* watering the trees unless they are located directly in the garden and receive water indirectly when the vegetables are watered."[32] To say that women are "not good at" watering trees is to sidestep the blunt reality, which is that they often simply refuse to water trees located outside of their gardens when they do not directly benefit from doing so. This has led forestry planners to conclude that, if women gardeners are to prove a reliable source of labor for orchard development, the task of watering landholders' tree seedlings must be imposed on them.

The stark clarity of this policy of labor co-optation underscores the fact that female gardeners are not the supposed "environmental altruists" ecofeminist writers sometimes make them out to be.[33] This is not to say, however, that women have no appreciation of the economic and ecological significance of garden-based tree planting. To the contrary, close inspection reveals that gardeners often plant trees in their garden plots, both as a source of additional income and as a way to produce microenvironmental conditions more favorable to vegetable production.[34] A systematic review of the competing sets of interests held by gardeners, landholders, and developers vis-à-vis agroforestry is thus necessary in order to fully assess the political fallout of Gambian environmental stabilization efforts.

The Meanings of Trees: Land, Labor, and Stabilization

The Tree Tenure Question

In the customary land law practiced among the Mandinka speakers who inhabit most of the rural garden districts on The Gambia's North Bank, a basic distinction is maintained between upland areas and swamplands. Since the virtual demise of the female-grown *Digitaria exilis* (Mandinka: *findo*) grain crop in the 1940s,[35] uplands in the area have been a fairly exclusive male production domain devoted to groundnut and coarse grain production; swamps, conversely, have been used for rice production by women.[36] The primary distinction between these two zones is that upland plots are only heritable through patrilineal social structures, whereas swamplands are heritable through both male and female kin. The authority to allocate use rights is also unevenly distributed: senior males are in a position to allocate rights to both uplands and swamps, whereas the prerogatives of senior women are restricted to rice-growing areas.

Virtually all of the twenty-odd communal garden sites considered in this research have been founded on patrilineal lands. Each site originated with a usufructuary land grant from one or more of the relatively senior men in the area's founding lineages. The gardens are typically designated according to the male landholder's name, and he maintains a certain degree of authority with respect to decisions governing land-use practices within the designated parcel. This may include determination of which crops are to be grown, when, and by whom, and whether specific forms of land development such as well construction or fencing may take place. Three general principles of tree tenure incorporated within Mandinka customary law have special relevance for garden lands.[37] First, tree-planting and land-use rights are partible; trees belong to those who plant them and not necessarily to those whose lands are used for tree-planting purposes.[38] Second, tenure rights frequently vary according to tree species, with ownership being most clearly articulated in the case of nonnative or introduced species. And third, the rights to tree benefits are traditionally somewhat diffuse. As Osborn explains:

> Benefit distribution is dominated by an owner's obligations to their family and community. In this regard, the concept of property as a right not to be excluded . . . is important. . . . Although the link between tree planting and tree ownership is direct and strong, the distinction of which *individual* owns a tree is less important than which *compound* [family-based residential unit] the tree belongs to.[39]

Given these guidelines, it is clear that the practice of tree planting in garden plots raises several challenges to the prevailing system. In principle, women are allowed to plant trees. A critical question, however, is whether the relatively weak usufructuary claims most women hold to garden lands embrace discretionary control over tree planting on those locations. Unilateral decisions by women to plant trees in the interstices of the prevailing upland/lowland land-use dichotomy could represent an erosion of patriarchal land-use controls. The promotion of nonnative fruit tree species for cultivation marks a second shift, insofar as it has encouraged privatization of orchards and a tightening of tree tenure throughout the country.[40] This, in turn, has undermined the third principle, according to which those who plant trees are expected to share benefits somewhat liberally among their affines. Landholders rhetorically justify orchard developments as wise investments serving all members of their lineages, but pocket earnings from fruit sales as private income nevertheless.

Gardeners' Perspectives

The prospect of solidifying usufruct rights through tree planting would seem to offer a significant advantage to women gardeners, yet their attitudes toward agroforestry remain decidedly mixed. A primary rationale for women engaging in tree planting involves the income derived from garden projects. The economic circumstances surrounding the garden boom have been such that families have had to rely more and more heavily on irrigated production in garden perimeters to ensure social reproduction from one year to the next. Gardeners have assumed unprecedented responsibilities for providing food, purchasing clothing, paying for ceremonial expenses, and meeting a wide array of other financial needs, and there is considerable social pressure for them to continue to do so.[41] At the same time, the cash income from vegetable sales has given women significantly greater autonomy, emboldening them to seek out additional sources of wealth accumulation. Fruit trees figure prominently in these strategies because they increase garden income and extend the seasonal pattern of earnings beyond the peak gardening period. The timing of crop sales is especially significant. Papayas and oranges, for example, mature during periods when production of other vegetables in the garden is low, and when demand for cash for food and ceremonial expenses is high. Strategic use of cash reserves during these periods of financial stress helps women "buy" freedom to engage more intensively in vegetable cultivation during the garden season.[42]

Returns per hectare from fruit and vegetable production, however, increasingly favor vegetables, even as shade problems from trees planted in gardens intensify. Consequently, while growers in some areas continue planting trees, others have begun cutting them back or chopping them down in order to open up the shade canopy and expose their vegetables to sunlight. As one women's group leader put it: "We are afraid of trees now. . . . You can have one [vegetables or fruit] or you can have the other, but you can't have both." The issue is thus not simply a question of rights to *plant* trees, but also of whether gardeners maintain control over decisions affecting the selection of species, the location of trees, and rights of trimming or removal, which is to say, the substance of property rights and the labor process. Women have demonstrated their capacity to manage fruit and vegetable crop mixes on their own account, but such control is called into question when landholders themselves choose to establish orchards on garden sites.

Landholders' Perspectives

In addition to income generation, the primary motivations men bring to tree planting in gardens include regaining lost social prestige and overcoming perceived threats to the lineage landholding system. As noted earlier, the generally poor groundnut market has left male farmers searching for ways to diversify their economic activities. They view the surge in female incomes as having created an imbalance in social relations that needs to be redressed. According to this perspective, women gardeners have lost respect for men, and their newfound wealth has left them in the position of being able to buy themselves out of "bad" marriages altogether. Tree planting accordingly has a double significance for landholders: it is a means of claiming the gendered subsidies that are embodied in fences and wells constructed with WID money in the early 1980s; and it constitutes a reassertion of male landholding privilege. In this context, all investments, whether from WID programs or forestry projects, are welcome, since landholders stand to benefit from them in the long run anyway. The key is the enclosure and bounding of female labor resources that make private orchard development possible.

Developers' Perspectives

As male orchard owners and female gardeners compete for land and labor, environmentalists bent on stabilization have intervened to tip the balance in favor of men. The actions of many agencies involved in tree planting threaten to undermine a female livelihood strategy the agencies themselves once supported. They have also evidently prompted isolated acts of resistance on the part of women in the form of tree seedling sabotage.[43] On the face of it, then, continued emphasis on garden siting is somewhat curious. One survey of land use undertaken by foresters grouped gardens with housing, roads, and other "communal" areas in a single land-use category, yet these "communal" uses still accounted for only 2 percent or less of all land area in the villages surveyed.[44] The fact that the environmental impact of gardens/orchards is so small begs the question of what really lies behind the movement to create them in the first place.

Developers have come under increasing pressure from various constituencies to demonstrate that they are doing something to quell the "eco-disaster" looming in the Third World.[45] It is for this reason that the relatively tiny, oasis-like women's gardens have assumed larger-than-life proportions. Reconstructed by developers as orchards, they offer up something of a mirage of ideological commitment, action, and success on an otherwise fairly bleak horizon, and come to form the basis of the developers' institutional legitimacy vis-à-vis their respective donor communities.[46] These developments constitute no mirage for vegetable growers, however. Instead, the coherence of policy statements and practice bespeaks the formation of a new political alliance in The Gambia's garden districts. Whereas the coalition confronting landholders in the 1970s and early 1980s sought to shore up the property claims of fledgling horticultural groups vis-à-vis the entrenched interests of the lineage-based landholding system, the late 1980s and the 1990s have seen developers effectively switch sides and endorse landholders as they move to reestablish prerogatives over garden plots and implement plans for income generation. One orchard manager engaged in a group venture backed by a development agency conveyed a clear sense of the new social dynamics of gardens/orchards: "We [project management] will take the advice of agriculture,[47] not the women. . . . The garden will not belong to them, but will belong to us. . . . The [women's] plots will not be permanent; [they] will be temporary. . . . [Management will make sure that the women] take great care of trees; . . . the survival of trees is their first goal."[48] This single-minded focus on trees and tree crop commodities grows directly out of the general ethos of the stabilization-cum-commercialization approach. The net effect has been disruption of a shared-access system of tree benefit distribution and the undermining of a relatively lucrative and socially beneficial market garden sector, all in the interest of promoting privatized orchard production with a limited environmental impact.

Conclusion

As Peluso notes: "Foresters tend to think of themselves as neutral experts carrying out their science according to the state's will; they rarely view either their own policies or their implementation methods as 'political' acts. From the forest dweller's point of view, however, nothing could be further from the truth."[49]

The economic barriers thrown up by the slow processes of rejuvenation and repair often preclude the prospect of environmental programs paying for themselves outright. Most individual and many private corporate actors are therefore reluctant to undertake the tasks of reclamation unilaterally. Instead, they default to state functionaries (or their surrogates—environmentalists), who confront the production obstacles on their behalf. The state can often alleviate some of the financial burdens of reclamation efforts through regressive taxation or unequal exchange mechanisms, but these tactics have their political and economic limits. Alternatively, state managers can opt for approaches involving coercion or regulatory controls, but these too can be quite expensive and are often ineffective because of the political resistance they engender. Consequently, environmentalists develop approaches driven by the positive incentive of profit taking associated with commodity production. These steps are intended to entice resource managers to undertake (or resume) stabilizing practices voluntarily, with only minimal outside financial assistance.

Commodity incentives notwithstanding, environmental entrepreneurs continue to resolve production dilemmas through dubious forms of social control, a point that is illustrated quite clearly in the recent history of Gambian agroforestry efforts. Village woodlots were originally conceived as a means to speed up the regeneration of firewood and building materials for rural Gambians. They ultimately failed on their own terms because villagers could not be enticed into providing labor under conditions where equitable distribution of benefits was neither guaranteed nor directly forthcoming. The shift in emphasis to fruit tree species that brought quicker returns helped resolve some of the contradictions that blocked the woodlot program, but garden/orchard managers were forced to co-opt an otherwise unwilling female labor force in the process.

In sum, although the drive to follow the commodity road grows out of recognition that environmental conditions are declining in many areas, realization that the economics of overcoming natural barriers to stabilization are problematic, and a desire to avoid the negative aspects of coercive and regulatory policies, it is also just as clearly born of economic opportunism in the face of crisis.[50] Blaikie and Brookfield's aphorism that "one person's degradation is another's accumulation"[51] can accordingly be read as a double entendre linking environmental decline to private gains derived as a consequence of environmental intervention.

Given the apparent ecological problems in many parts of the world, and the routine failure, not to mention political shortcomings, of coercive and regulatory practices to produce desired changes in land-use practices, the investigation of alternative roads to environmental stabilization is warranted. These concerns notwithstanding, environmentalists who would embrace the commodity road as a form of radical environmental practice must contend with the fact that the commodity road is itself often fraught with contradictions. Evidence from The Gambia suggests that the commodification of nature can lead to the imposition of new forms of property claim and the introduction of inequitable labor relations. Although the socially necessary tasks of environmental reproduction, protection, and restoration are critical, dispossession and exploitation in the name of a "common" environmental good cannot be justified.

Notes

1. Piers Blaikie and Harold Brookfield, *Land Degradation and Society* (London: Methuen, 1987), 14.
2. The author would like to acknowledge the following funding sources for generous support during fieldwork: the Fulbright-Hays Doctoral Dissertation Research Award, the Social Science Research Council/American Council of Learned Societies International Doctoral Research Fellowship for Africa, the Rocca Memorial Scholarship for Advanced African Studies, and the National Science Foundation Fellowship in Geography and Regional Science. Special thanks to Michael Watts, Rod Neumann, Dorothy Hodgson, Susanna Hecht, Saul Halfon, Oliver Coomes, Andrew Stewart, and Neil Smith, who read the manuscript and offered advice and encouragement. The original version of this chapter was published in *Antipode* 27, no. 4 (1995): 325–42. I am grateful to the editors for their permission to reprint it here.
3. Kellman draws a useful distinction between "avoidance" and "control" approaches. Land "set-asides" and nature preserves are "avoidance" strategies, whereas more direct and immediate steps toward slope stabilization, water channeling and retention, and adjustment of nutrient levels in soils imply human "control" (M. Kellman, "Some Implications of Biotic Interactions for Sustained Tropical Agriculture," *Proceedings of the Association of American Geographers* 6 [1974]: 142–45; cited in Blaikie and Brookfield, *Land Degradation and Society*, 8).
4. Blaikie and Brookfield, *Land Degradation and Society*. Nancy Peluso illustrates the distinctive economic problems of reclamation in her study of Javanese forests: "profits, wages, commissions, and bribes came from *cutting* big trees, not replacing them. Replacing them, indeed, required paying day laborers hired for each separate task of clearing brush, planting, and weeding" (Nancy Peluso, *Rich Forests and Poor People: Resource Control and Resistance in Java* [Berkeley: University of California Press, 1992], 63; emphasis added). She also notes that the problematic economics of reforestation cannot be remedied by mechanization.
5. Blaikie and Brookfield, *Land Degradation and Society*, 245.
6. Ibid., 245–46. See Nancy Lee Peluso, "Coercing Conservation? The Politics of State Resource Control," *Global Environmental Change* 3, no. 2 (1993): 199–217.
7. Raymond Bonner, *At the Hand of Man: Peril and Hope for Africa's Wildlife* (New York: Vintage Books, 1993); Peluso, *Rich Forests and Poor People*; Peluso, "Coercing Conservation?"; Roderick P. Neumann, "Ways of Seeing Africa: Colonial Recasting of African Society and Landscape in Serengeti National Park," *Ecumene* 2, no. 2 (1995): 149–69; Jonathan Adams and Thomas McShane, *The Myth of Wild Africa* (New York: Norton, 1992); Susanna Hecht and Alexander Cockburn, *The Fate of the Forest: Devel-*

opers, *Destroyers and Defenders of the Amazon* (New York: HarperCollins, 1990); William Beinart, "Soil Erosion, Conservationism and Ideas about Development: A Southern African Exploration, 1900–1960," *Journal of Southern African Studies* 11, no. 1 (1984): 52–83.

8. Hecht and Cockburn, *The Fate of the Forest*; Anthony Anderson, ed., *Alternatives to Deforestation: Steps toward Sustainable Use of the Amazon Rain Forest* (New York: Columbia University Press, 1990); Edward Whitesell, "Future Grassroots Reserves in the Amazon Basin: Alternative Landscapes of Conservation," in *Proceedings, 48th International Congress of Americanists* (Stockholm: Institute of Latin American Studies, Stockholm University, 1994).
9. For example, by the forest products or livestock ranching industries; see Susanna Hecht, "Environment, Development and Politics: Capital Accumulation and the Livestock Sector in Eastern Amazonia," *World Development* 13, no. 6 (1986): 663–84; Brent Millikan, "Tropical Deforestation, Land Degradation, and Society: Lessons from Rondonia, Brazil," *Latin American Perspectives* 19, no. 1 (1992): 45–72.
10. Jason Clay, *Indigenous Peoples and Tropical Forests: Models of Land Use and Management from Latin America* (Cambridge, Mass.: Cultural Survival, 1988); Alan Durning, *Guardians of the Land: Indigenous Peoples and the Health of the Earth*, Worldwatch Paper 112 (Washington, D.C.: Worldwatch Institute, 1992); cf. Jonathan Friedmann and Haripriya Rangan, eds., *In Defense of Livelihood: Comparative Studies on Environmental Action* (West Hartford, Conn.: Kumarian Press, 1993).
11. Roderick P. Neumann, "Local Challenges to Global Agendas: Conservation, Economic Liberalization and the Pastoralists' Rights Movement in Tanzania," *Antipode* 27, no. 4 (1995): 363–82.
12. Vandana Shiva, *Staying Alive: Women, Ecology and Development* (London: Zed Press, 1988); Richard Schroeder and Krisnawati Suryanata, "Gender and Class Power in Agroforestry: Case Studies from Indonesia and West Africa," in *Liberation Ecologies: Environment Development Social Movement*, ed. Richard Peet and Michael Watts (London: Routledge, 1996).
13. Richard Peet and Michael Watts, "Development Theory and Environment in an Age of Market Triumphalism," *Economic Geography* 69, no. 3 (1993): 227–53; Michael Watts, "Development II: Markets, States and the Privatization of Everything," *Progress in Human Geography* 18, no. 3 (1994): 371–84.
14. See Judith Carney, "Struggles over Crop Rights and Labour within Contract Farming Households in a Gambian Irrigated Rice Project," *Journal of Peasant Studies* 15, no. 3 (1988): 334–49; Judith Carney, "Struggles over Land and Crops in an Irrigated Rice Scheme: The Gambia," in *Agriculture, Women and Land*, ed. Jean Davison (Boulder, Colo.: Westview Press, 1988), 59–78; Judith Carney, "Peasant Women and Economic Transformation in The Gambia," *Development and Change* 23, no. 2 (1992): 67–90; Judith Carney and Michael Watts, "Disciplining Women? Rice, Mechanization and the Evolution of Gender Relations in Senegambia," *Signs* 16, no. 4 (1991): 651–81; Jennie Dey, "Gambian Women: Unequal Partners in Rice Development Projects?" *Journal of Development Studies* 17, no. 3 (1981): 109–22; Jennie Dey, "Development Planning in The Gambia: The Gap between Planners' and Farmers' Perceptions, Expectations and Objectives," *World Development* 10, no. 5 (1982): 377–96; and Patrick Webb, "Of Rice and Men: The Story behind The Gambia's Decision to Dam Its River," in *The Social and Environmental Effects of Large Dams*, vol. 2, ed. Edward Goldsmith and Nicholas Hildyard (Wadebridge, U.K.: Wadebridge Ecological Centre, 1984), 120–30.
15. Government of The Gambia, *Executive Summary: Programme for Sustained Development: Sectoral Strategies* (Banjul: Gambia Round Table Conference, n.d.); Government of The Gambia, "National Natural Resource Policy" (Banjul: Government of

The Gambia, 1990); United States Agency for International Development (USAID), *Agricultural and Natural Resource Program: Program Assistance Initial Proposal* (Banjul: USAID, 1991); Wolfgang Thoma, *Possibilities of Introducing Community Forestry in The Gambia*, Part I (Banjul: Gambia-German Forestry Project, Deutsche Gesellschaft für Technische Zusammenarbeit [GTZ], 1989); Agroprogress International, "Project Preparation Consultancy for an Integrated Development Programme [European Community] for the North Bank Division" (Bonn: Agroprogress International, 1990); Arthur Thiesen, Saidu Jallow, John Nittler, and Dominique Philippon, "African Food Systems Initiative, Project Document" (Banjul: U.S. Peace Corps, 1989); Law Reform Commission of The Gambia, *The Customary Laws of The Gambia* (Banjul: Law Reform Commission of The Gambia, 1988); Judith Carney, "Converting the Wetlands, Engendering the Environment: The Intersection of Gender with Agrarian Change in The Gambia," *Economic Geography* 69, no. 4 (1993): 329–48.

16. In 1990, the maternal mortality rate in The Gambia was 10 per 1,000 live births, with some areas reporting rates as high as 20–22; infant and child mortality stood at 143 and 242 per 1,000, respectively, and the overall fertility rate was 6.5 live births per woman (data provided by the World Bank WID project in The Gambia). These rates are all among the highest in Africa (Kevin Cleaver and Götz Schreiber, *The Population, Agriculture and Environment Nexus in Sub-Saharan Africa* [Washington, D.C.: World Bank, 1993]).
17. Richard Schroeder and Michael Watts, "Struggling over Strategies, Fighting over Food: Adjusting to Food Commercialization among Mandinka Peasants in The Gambia," in *Research in Rural Sociology and Development*, vol. 5, *Household Strategies*, ed. Harry Schwarzweller and Daniel Clay (Greenwich, Conn.: JAI Press, 1991), 45–72.
18. A study conducted by UNICEF consultants in neighboring Senegal suggests that the caloric requirements of manually irrigating garden plots may outweigh nutritional benefits gained through the improved availability of fruits and vegetables (Karen Schoonmaker-Freudenberger, "L'intégration en faveur des femmes et des enfants: une évaluation des projets régionaux intégrés soutenus par le Gouvernement du Sénégal et UNICEF" (Dakar: Government of Senegal and UNICEF, 1991). Similar unofficial results were reported by a researcher in The Gambia's maternal and child health-care program to a meeting of the Gambian Food and Nutrition Association (GAFNA) in 1986. On the sale of vegetables as a primary source of cash income, see Richard Schroeder, "Shady Practice: Gender and the Political Ecology of Resource Stabilization in Gambian Garden/Orchards," *Economic Geography* 69, no. 4 (1993): 349–65.
19. Cathy Jabara, *Economic Reform and Poverty in The Gambia: A Survey of Pre- and Post-ERP Experience*, Monograph 8 (Ithaca, N.Y.: Cornell Food and Nutrition Policy Program, 1990); Joachim von Braun, Ken Johm, Sambou Kinteh, and Detlev Puetz, *Structural Adjustment, Agriculture and Nutrition: Policy Options in The Gambia*, Working Papers on Commercialization of Agriculture and Nutrition, no. 4 (Washington, D.C.: International Food Policy Research Institute [IFPRI], 1990). The garden boom has taken place against the backdrop of a downward trend in annual precipitation. Norton et al., report that, during a twenty-year period prior to 1989, average rainfall and the length of the rainy season declined 24 percent to 36 percent, and by fourteen to twenty-four days, respectively, in different parts of the country (George Norton, Brad Mills, Elon Gilbert, M. S. Sompo-Ceesay, and John Rowe, "Analysis of Agricultural Research Priorities in The Gambia," unpublished manuscript [Banjul: Department of Agricultural Research, Ministry of Agriculture, 1989]); see also Robert Mann "Africa: The Urgent Need for Tree-Planting," unpublished manuscript (Brikama, The Gambia: Methodist Church Overseas Division, 1989).
20. Government of The Gambia, "National Natural Resource Policy" (Banjul: Govern-

ment of The Gambia, 1990); see also Thoma, *Possibilities of Introducing Community Forestry in The Gambia*, Part I; E. Kennel, "Vegetation and Land Use Inventory: Change of Forest Vegetation Cover from 1980 to 1988," report no. 22 (Banjul: Gambian-German Forestry Project, 1990).

21. Banjul Declaration (Banjul: Government of The Gambia, February 18, 1977).
22. Thoma, *Possibilities of Introducing Community Forestry in The Gambia*, Part I.
23. Ibid., 37; see also Steven Lawry, "Report on Land Tenure Center Mission to The Gambia" (Madison: University of Wisconsin Land Tenure Center, 1988).
24. Thoma, *Possibilities of Introducing Community Forestry in The Gambia*, Part I, 44; see also Sarah Norton-Staal, "Women and Their Role in the Agriculture and Natural Resource Sector in The Gambia" (Banjul: USAID, 1991); Mann, "Africa: The Urgent Need for Tree-planting"; Lawry, "Report on Land Tenure Center Mission to The Gambia."
25. Lawry, "Report on Land Tenure Center Mission to The Gambia," 2–3. Cf. John Raintree, ed., *Land, Trees and Tenure: Proceedings of an International Workshop on Tenure Issues in Agroforestry* (Madison: University of Wisconsin Land Tenure Center, 1987); Louise Fortmann and John Bruce, eds., *Whose Trees? Proprietary Dimensions of Forestry* (Boulder, Colo.: Westview Press, 1988).
26. Lawry, "Report on Land Tenure Center Mission to The Gambia," 2–3.
27. This arrangement closely resembles the *taungya* system, a type of agroforestry that "is initially dominated by the agricultural component (one to three years) but taken over by the forestry-tree components for the long term (twenty years or more)" (Peluso, *Rich Forests and Poor People*, 146), with the rights to end products shifting from farmers to tree-crop entrepreneurs during the latter stage of its development. The large *Gmelina arborea* plantation in Lower River Division on The Gambia's South Bank was established under the colonial government in 1959 in a similar manner. When the *Gmelina* seedlings were first planted, farmers were granted privileges to plant groundnuts on the site between the rows of seedlings for one year. Land was free of charge, provided the farmers tended the tree seedlings. Subsequent efforts to establish teak forests under the same *taungya* principles were frustrated, however, when farmers neglected and/or destroyed tree seedlings in order to prolong access to land for groundnut production (Thoma, *Possibilities of Introducing Community Forestry in The Gambia*, Part I). For an incisive analysis of the origins of *taungya* practices in colonial Burma, see Raymond Bryant, "The Rise and Fall of *Taungya* Forestry: Social Forestry in Defence of the Empire," *Ecologist* 24, no. 1 (1994): 21–26.
28. Cited in Norton-Staal, "Women and Their Role in the Agriculture and Natural Resource Sector in The Gambia," 12.
29. Worldview International Foundation (WIF), *WIF Newsletter* 3, no. 1 (1990): 6; emphasis added.
30. USAID, "Agricultural and Natural Resource Program, Program Assistance Initial Proposal" (Banjul: USAID, 1991), 28.
31. See Bina Agarwal, "The Gender and Environment Debate: Lessons from India," *Feminist Studies* 18, no. 1 (1992): 119–58; Dianne Rocheleau, "Gender, Ecology, and the Science of Survival: Stories and Lessons from Kenya," *Agriculture and Human Values* 8 (1991): 156–65; Fiona MacKenzie, "Exploring the Connections: Structural Adjustment, Gender and the Environment," *Geoforum* 24, no. 1 (1993): 71–87; Rosi Braidotti, Ewa Charkiewicz, Sabine Häusler, and Saskia Wieringa, *Women, the Environment and Sustainable Development: Towards a Theoretical Synthesis* (Atlantic Highlands, N.J.: Zed Books, 1994).
32. Quoted in Norton-Staal, "Women and Their Role in the Agriculture and Natural Resource Sector in The Gambia," 6; emphasis added.

33. The tendency to naturalize the connection between women and their environments has been the subject of considerable debate within the feminist community. See Janet Biehl, *Rethinking Ecofeminist Politics* (Boston: South End Press, 1991); Carolyn Merchant, *Radical Ecology: The Search for a Livable World* (New York: Routledge, 1992); Braidotti et al., *Women, the Environment and Sustainable Development.*
34. Hazel Barrett and Angela Browne, "Environmental and Economic Sustainability: Women's Horticultural Production in the Gambia," *Geography* 76, no. 332 (1991): 241–48; Mann, "Africa: The Urgent Need for Tree-Planting."
35. David Gamble, *Economic Conditions in Two Mandinka Villages: Kerewan and Keneba* (London: Colonial Office, 1955).
36. Large-scale irrigated rice-growing perimeters in central Gambia, which male heads of household control, are exceptions to this general pattern (Carney, "Struggles over Crop Rights and Labour within Contract Farming Households"; Carney, "Struggles over Land and Crops in an Irrigated Rice Scheme").
37. Elizabeth Osborn, "Tree Tenure: The Distribution of Rights and Responsibilities in Two Mandinka Villages" (M.S. thesis, University of California-Berkeley, 1989).
38. This practice is not uncommon elsewhere (Raintree, *Land, Trees and Tenure*; Fortmann and Bruce, *Whose Trees?*).
39. Osborn, "Tree Tenure," 66; emphasis added.
40. Schroeder, "Shady Practice."
41. Ibid.
42. Richard Schroeder, "'Gone to Their Second Husbands': Marital Metaphors and Conjugal Contracts in The Gambia's Female Garden Sector," *Canadian Journal of African Studies* (forthcoming). Vegetable production in The Gambia takes place during the dry season (November–June), a period of slack labor demand when men's expectations of household services (e.g., the quality of food preparation) are high. Vegetable growers have had to work hard at winning their husbands' support, given the time-consuming nature of their horticultural activities (Hazel Barrett and Angela Browne, "Time for Development? The Case of Women's Horticultural Schemes in Rural Gambia," *Scottish Geographical Magazine* 105, no. 1 [1989]: 4–11).
43. Schroeder, "Shady Practice."
44. Wolfgang Thoma and S. Jaiteh, *Possibilities of Introducing Community Forestry in The Gambia*, Part III (Banjul: Gambia-German Forestry Project, Deutsche Gesellschaft für Technische Zusammenarbeit [GTZ], 1991).
45. Jonathan Barker, *Rural Communities under Stress: Peasant Farmers and the State in Africa* (Cambridge: Cambridge University Press, 1989); Michael Watts, "The Agrarian Crisis in Africa: Debating the Crisis," *Progress in Human Geography* 13, no. 1 (1989): 1–43.
46. Schroeder, "Shady Practice."
47. The reference here is to the Department of Agriculture's horticulture extension agents.
48. Interview with the author, April 1991.
49. Peluso, *Rich Forests and Poor People*, 8.
50. Watts, "The Agrarian Crisis in Africa."
51. Blaikie and Brookfield, *Land Degradation and Society*, 14.

Discipline or Solidarity? Ecology as Politics

Yrjö Haila

Introduction: The Multiple Nature of Ecological Catastrophes

One of the reactors in the nuclear power plant at Chernobyl, Ukraine (then part of the Soviet Union), exploded in April 1986. The event has been characterized innumerable times as an "ecological catastrophe." But what is the "ecology" that was catastrophically affected by the explosion?

The answer is not immediately obvious. Since the accident, ecologists have worked in the area around Chernobyl tracing its consequences on natural ecological systems. Immediate visible damage was recorded to coniferous trees in an area of thirty square kilometers surrounding the reactor site, and some damage within a radius of thirty kilometers around the site. Corresponding figures have been obtained in studies of soil organisms.[1] Thus, although it is possible that some longer-term effects have gone unnoticed at greater distances, the event definitely was not a "catastrophe" if the immediate effect on natural ecological systems is taken as a criterion.

The "catastrophic" dimension stems instead from *human* suffering caused by the accident. This is due, as is well known, to radiation, and is mediated to future generations by hereditary mutations. Reasonable estimates of health hazards, expressed as an estimated number of deaths by cancer caused by the accident, vary between several thousand and several tens of thousands. The estimates are uncertain because the cases are distributed over a long period of time and because people also get cancer without being exposed to additional doses of radiation. Consequently, even more important than direct health hazards is the stress *of uncertainty* among people who were exposed to radiation. For instance, among five thousand Estonians, mainly conscripts, sent to Chernobyl to help with the cleanup after the accident, the death rate caused by serious illnesses was lower during the following four years than in a control group among the rest of the Estonian population (the physical fitness of conscripts is above average), but the suicide rate was slightly higher, and a clearly higher proportion suffered from psychosocial problems.[2]

Is human suffering an ecological catastrophe? The explosion at Chernobyl presents us with only minor effects on nature "out there" but with increasingly serious consequences when the focus is shifted from the environment to the physical health of people and to their psychological and social well-being. These consequences are the result of radiation, and they are mediated through ecological processes such as food-web relationships that cause the enrichment of radioactive elements in the environment of people and ultimately in their bodies. *The ecology on which human beings depend was destroyed*: the accident was an ecological catastrophe when ecology is adequately defined.

This brings into focus the "ambiguity of Chernobyl": the "ecology" primarily damaged by the accident was not "out there" but *within the human body and society*. The ambiguity is corroborated by what is known about the effects of radioactive fallout in the atmosphere following nuclear tests. The doses of radioactivity released into the atmosphere have been orders of magnitude larger than what dispersed from Chernobyl but, again, now that details of the consequences are slowly emerging, it is human illness and suffering that dominate the picture, not disruption of natural ecosystems (except, of course, on the actual test sites). This is true concerning each one of the test sites from the Pacific atolls to Maralinga (Australia), Novaya Zemlya (Russian arctic), Semipalatinsk (Kazakhstan), and the Great Basin deserts in Utah and Nevada.[3]

Chernobyl demonstrates the multiple character of the "ecological crisis." The notion refers to all human-caused threats to ecological systems of the earth put together, but the threats drawing most attention concern primarily human health and well-being. It is ultimately impossible to distinguish from each other conditions controlling ecological systems and conditions controlling human existence, although different aspects of this relationship can be isolated for specific analytic purposes. A major issue in each particular case is scaling, that is, defining the spatial and temporal extent of the effects studied. Intensive impacts that are spatially and temporally restricted have quite different consequences than do weak impacts that extend over a large area and long period of time.[4] Nevertheless, it will always turn out that what was supposed to be strictly a deterioration of nature also concerns human well-being, and what was supposed to be purely a problem of human well-being also relates to the human environment, that is, human ecology. It is therefore misleading to postulate a strict boundary between society and nature.

The ecological crisis is a shock of uncertainty. It is the realization that although there are many potential Chernobyls around, it is impossible to identify them with certainty in advance. Human existence depends on the life-support systems of the earth and there can be no transcendental guarantees that those systems are not disrupted by human activity. Immediate and local processes such as pollution of a lake, depletion of a fish stock, or erosion of a hillside are readily visible. But is the pollution, or depletion, or erosion going on here and now a universal trend out of human control or is it, in contrast, a local aberration that is easily controlled? This question seldom has a definite answer. Such an uncertainty about the ecological sustenance of human existence tends to turn into certainty, that is, certainty about risks—a risk consciousness that never fades away.[5]

There cannot be any ultimate solution to the ecological crisis. How to go about trying to solve an unsolvable problem? One common technique is to construct the problem in such a way that it appears solvable. In this essay, I suggest that ideas about *social order* play a constitutive role in the phrasing and rephrasing of ecological problems in such a way that they appear solvable. Order is the opposite of chaos; it is a prerequisite for governability, controllability, tractability. Thus, it may easily seem that there is no alternative to assuming that order in both nature and society is a prerequisite for coping with the relationship between humanity and nature. But, as human existence ultimately depends on the life-support systems of the earth, order in society is brought into the focus. My thesis is that a particular conception of social order is inherent in most definitions of ecological problems and in suggestions on how to manage those problems. I call this the principle "social order first."

Ecological Problems and Social Order

Being a shock of uncertainty, the ecological crisis has implications for all areas of human activity. Consequently, "the environment" cannot be isolated as a specific field of policy. This conclusion brings forth the following tension: is it possible to just start "solving the problems," or is it necessary first to "create (secure) social conditions such that the problems can be solved"? Is it possible even to imagine a solution without making implicit assumptions about a state of society that would make those solutions possible?

A view derived from traditional revolutionary ideology regards the socialist revolution as a necessary condition for adequately addressing ecological problems. This, as a matter of fact, is one reason why traditional Marxists have been slow to take up ecological issues. A conservative variant emphasizes economic health as a basic precondition for any kind of social action. However, the argument about social order has another foundation that has considerable appeal for ordinary sound reason: to achieve anything at all, the governability and order of society need to be maintained.

Human societies have been threatened by dissolution caused by internal conflicts as long as they have existed.[6] Efforts and procedures to cope with this threat have, however, changed throughout history. The goal of maintaining social order is historically concrete and always tied to a particular system of power. In the modern world, the genealogy of the question of social order goes back to the gradual establishment of great territorial, centralized, and administratively unified states. This was accompanied by the opposite tendency toward increasing autonomy of individual citizens liberating themselves from subjugation to traditional religious authority and earthly sovereignty. This general picture was painted by the social theorists of the nineteenth century. According to Michel Foucault, government became gradually a "general problem" during the transition from traditional sovereignty to modern state: a need was perceived to establish unified principles of governmental conduct reaching from individual self-government through the economy of the family to the ruling of the state.[7] Foucault called this requirement "governmentality" and gave it the following characterization: "The ensemble formed by the institutions, procedures, analyses and reflections, the calculations and tactics that allow the exercise of this very specific albeit complex form of power, which has as its target population, as its principal form of knowledge political economy, and as its essential technical means apparatuses of security."[8]

Foucault points to the deep internal connection between problems the government is supposed to address ("population" in all its various manifestations) on the one hand, and the forms of knowledge and administration the government has at its disposal on the other hand. In other words, the very logic of "governmentality" presupposes a commitment to prevailing forms of social order and apparatuses maintaining them. One can expect that management of any particular problem area, such as the environment, is subjected to the same logic.

Simultaneously with the rise of government as a general problem, a shift occurred in the role of science and in the view of the relationship of society to nature. Zygmunt Bauman has argued that the shift was backed by a totalizing ethos starting from the premise that society as a whole can be organized in a completely rational way, according to a unified grand scheme. This intellectual legislation, or "gardening," was brought to fruition by the philosophers of the French Enlightenment. Nature was, in this view, entirely subordinate to human aims and goals; it was a law-governed collection of resources for fulfilling human ends.[9] This strengthens the mechanisms through which nature becomes entrapped within the sphere of human governmental rationality.

This grand, optimistic ethos is, of course, in serious trouble nowadays, but the issue of "governmentality" remains. It thus seems a priori impossible to separate solutions to ecological problems from doctrines concerning social order. Such a separation is additionally blurred by ambiguity in the definition of the environment: what Marston Bates originally distinguished as the "perceived" and the "conceptual" dimensions of the environment are continuously mixed together.[10] This is clear, for instance, when the quality of the environment is assessed. Such an assessment leans necessarily on a priori criteria of what a good environment is like. But this means that there is no purely empirical way of recording threats to a good environment either—when perceived, such threats are always already within a framework of accepted criteria, that is, preconceptualized. In a "risk society" potential, invisible, but preconceptualized and hence prerecognized hazards are everywhere, and such risks are culturally real irrespective of whether they are "strictly scientifically" grounded or not.[11]

The preconceptualization of the environment is a social process, impregnated by social power relationships. Elements of the environment of any society are laden with symbolic significance, and what is recognized as an environmental problem is a derivative of how the environment is valued. Health hazards that concern industrial, mining, or plantation workers, or traffic arrangements that neglect pedestrian safety, have not been automatically recognized as environmental problems. Biases occur also in more subtle and "natural" forms that stem, for instance, from how land is allocated to various purposes through "natural" economic processes. The production of space is simultaneously production of social control.[12] People living in a particular environment come across a preexisting structure that apparently is dictated by nature, but that in fact is a product of a historical process.

The perception of ecological problems in public consciousness is also dependent on the seriousness of other perceived threats. Traditional threat perceptions, such as military invasion or natural catastrophe causing starvation or disease, were supplemented in the 1950s with nuclear war. The relative weight of environmental threats grew rapidly in the 1980s, and they were incorporated into war scenarios via the threat of "nuclear winter." However, during the recession of the early 1990s, environmental threats were again reduced to a lower status in the hierarchy in opinion polls in many European countries.[13]

Science gives, of course, tools for recognizing environmental threats, but the evidence provided by science is not indisputable. One reason for this is that ecological hazards ought to be recognized before evidence that would prove the case is available. In methodological terms, the demand of replication of results is not fulfilled by environmental hazards: we do not want to "repeat" environmental catastrophes, we want to avoid them.[14] This opens up space for maneuver for those who want to downplay the problems, as has happened, for instance, in controversies over climate change, but this is unavoidable. The normative context of environmental research is similar to that of the medical sciences: threats need to be recognized and eliminated before they materialize. Thus, the science that proves that environmental threats exist by detecting symptoms cannot provide the means for removing their causes.[15]

Thus, criteria as to what constitutes a good environment cannot be derived from nature, but must be defined by people. This is exactly where the commonsense argument in support of the maintenance of social order creeps in: secure existence of human subjects and their ability to act are diagnostic features of life in a good environment. Consequently, assumptions concerning the prevailing social order form a prism through which ecological issues are interpreted. This gives apparent coherence to problems that in fact might be idiosyncratic, heterogeneous, and very unevenly distributed across different societies and social classes within societies.[16] This also implies that disciplinary control of people's behavior to maintain social order is an inherent tendency in environmental policy. This is what I call "social order first."

In the next section, I review briefly how the principle "social order first" has influenced major variants of thinking about environmental issues (albeit not without internal tensions). In the following section, I use recent discussions on "environmental security" to highlight some of the contradictions to which the interpenetration of ecological problems and social order gives rise. "Environmental security" serves as a paradigmatic example of this mixing of perspectives by its ties to the military, an authoritarian institution that has the task of maintaining social order. I claim that giving the military a role in environmental policy strengthens a disciplinary view of ecological problems and is ultimately counterproductive. In the final section, I outline an alternative conception of political ecology.

Unproblematic Ecological Prescriptions: Naturalization, Linearization, Moralization

A series of important books published during the 1960s paved the way for the era of the ecological crisis. The themes they addressed were the finiteness of the earth and its resources, the interconnectedness of ecological processes, and the cumulative effects of pollutants on the environment and on human health. Traditional conservation concerns such as the loss of species and habitats were also emphasized, but with a more pessimistic tone than previously, mainly because of the perceived threat of a human population explosion.[17]

Many of the themes addressed in the 1960s first arose in the nineteenth century,[18] but they were now perceived as more and more tightly integrated with social processes. In other words, views of social order became integral elements in the new problem definitions. For instance, Paul Ehrlich's "Eco-Catastrophe"—one of the most widely publicized early warnings—was an explicit social dystopia.[19] In Ehrlich's scenario, ecocatastrophe was triggered by competition among the superpowers for political influence in underdeveloped countries. To achieve this they invented ever more efficient insecticides (note the connection to Rachel Carson's *Silent Spring*), which ultimately contaminated and killed the oceans. However, the intellectual shift from progressivist utopia to environmentalist dystopia was originally smaller than it might appear. Visions of a total catastrophe, of the "death of nature," continue the hubris of Enlightenment reason, albeit in an inverted way: although humanity cannot completely control nature, humanity can completely destroy nature.

Environmentalist dystopia extrapolated trends from the present to the future, leaving the social background intact. What happened can be characterized as *naturalization* of the prevailing social order. The "naturalistic" assumption behind the Ehrlichian dystopia is that the growth rate of human populations is geometric,[20] and is unaffected by social conditions. Thus, the environmentalist anti-utopia falls back into Malthusianism, which views population growth—in connection with other growth processes in which growth rate is in linear relationship with the number of people—per se as the ultimate problem. This idea found its most explicit formulation in Garrett Hardin's "lifeboat ethics."[21] As a matter of fact, however, human population growth rates have varied greatly across societies. The assumption of a constant growth rate is a prefabricated "fact" taken as a natural fact, and shifted as such to the realm of social prescriptions.[22] The tendency to disciplinarity is obvious and protofascist in extreme Malthusianism: "nature" sets the limits that people have to obey, and they have to be enforced if necessary.

Malthusianism amounts to naturalization of ecological problems: both "nature" in the form of specified environments, and "humans" as biological beings are reconstructed as natural givens, and the potential role of history in modifying such relationships is denied altogether. A corollary of this naturalization is the inability of conservationists to understand processes that determine the economic development of particular regions because of their tendency to derive it from the "carrying capacity" of the natural environment. John Muir considered beekeeping the only viable activity in southern California, where Los Angeles with its satellite suburbs nowadays spreads.[23] What has happened in southern California during the past century has produced a regional ecological crisis, but beekeeping is certainly not the only viable alternative. Local economic development and local ecological conditions are increasingly uncoupled from each other in the modern world, which makes derivation of the economic prospects of a particular region from local natural conditions impossible.

Malthusianism thus reified social order through naturalization.[24] But the environmental awakening also led to another line of thought in which the existing social order was accorded an unquestioned primacy. This occurred when environmental problems were conceptualized within the framework of established management rationality. Ecology was used to derive "ecological principles" of pollution control or resource management. This tendency is technocratic reformism. It triggered the birth of environmental policy (or policies—if the measures required were different for energy, industries, agriculture, forests, fisheries, pollution control, etc.). This was *linearization* of ecological problems: traditional types of sectoral policies were applied to environmental issues one at a time.[25]

Technocratic reformism arose from the progressivist optimism of the early twentieth century that viewed "scientific management" as a method to solve problems in resource exploitation. The key concept was the model of "sustainable yield," which simply combined harvesting (social activity) and resource productivity (natural process).[26] The original model proved to be too simplistic and was replaced with more complex, systems-theoretical approaches that tried to incorporate indirect ecological effects such as pollution or long-term resource depletion. In each case, ecology was translated into managerial schemes without giving serious consideration to the traditions and cultures that produce and reproduce the actual practices adopted by people in the situations studied—again, the prevailing social order was a given.[27]

Technocratic reformism is dominant in research on global change, as is shown by the following characterization of this research task: "Basic scientific research will provide much improved knowledge of human-induced change in the biosphere, the capacity of natural systems to absorb such change, and the ability of human societies to adjust or adapt their behaviour."[28] The symptomatic point in this definition is that it allocates to "basic scientific research" the responsibility to deal with the whole chain from physical processes ("human-induced change") to the ability of human societies to adjust. It seems as if the identification of human-induced changes in the biosphere would dictate the social measures necessary for coping with them, that is, define a new framework for giving disciplinary prescriptions.[29]

A priority in technocratic reformism is to maintain a society capable of acting and providing solutions to problems perceived as urgent. What is recognized as a "problem" worth serious consideration is another matter. In its most democratic version, technocratic reformism may lean on public opinion and consensus, even give a role to pressure groups such as conservation movements, whereas more autocratic versions consider only criteria dictated by assumed necessities of economic development, and a major concern is to get people to adapt to increased risks.[30]

The evolution of technocratic reformism can be presented as a series of cycles in which ecological concepts were adopted, one after another, and transformed into managerial vocabulary. "Sustainable yield" was an early managerial approach to exploitation of single natural resources. "Ecosystem management," derived from ecosystems ecology, was a dominant topic in the 1970s. In the 1980s there followed "conservation biology," which applies population ecology and genetics to the preservation of endangered animals and plants. The theme for the 1990s seems to be "biodiversity valuation," which allocates monetary values to natural species and ecosystems and applies neoclassical environmental economics to their management.[31] All these traditions have provided new technical tools for coping with the respective problems—as technological development always does—but critical questions need to be asked about the context in which they are applied.

Another type of pragmatic reformism alternative to technocratic "linearization" has been prevalent within conservation movements, namely, *moralization* of ecological problems. These organizations emphasize individual responsibility, which can be perceived as more malleable than social mechanisms. The background is quite understandable. The problems are considered so urgent that a basic imperative is to achieve practical success, without any consideration of wider social scenarios (which may not even be acknowledged as a legitimate issue). Traditional nature conservation abounds with organizations that mostly avoid taking stands on social issues, for instance, the World Wildlife Fund, an organization created for collecting money for conservation, particularly for acquiring land. Social luminaries such as Prince Bernhard of the Netherlands (until his reputation was spoiled by the Lockheed bribery scandals) have been recruited to head the organization. The dilemma confronting conservation movements in this context is analogous to the problem of charity: does it matter that corporations with the poorest record of environmental protection often dominate the list of donors?[32] Conservation movements have not been insensitive to the dilemma, which has resulted in an uneasy compromise between criticizing the prevailing power structure and yielding to its pressures.

The moralizing tendency in environmentalism is often hostile to technocratic solutions, regarding them as useless and ultimately conterproductive attempts to evade the moral reponsibility of people to respect and guard natural ecosystems. Trends toward disciplinarity stem in more radical environmentalist groups from the pressure to find a "correct" spiritual contact with nature; this is the totalitarianism of being right.[33] On the other hand, the organizations may be completely blindfolded relative to social power structures.

Within each variant of environmentalist thought there are internal tensions concerning the question, What is the relationship of the actual to the desired social order? This tension is scale-dependent. It is relatively easy to figure out general blueprints and harmonize ecological goals and social order with each other on the scale of the whole society, but this becomes more difficult concerning specific social practices that are controlled by their own, historically shaped rules and ethos. For instance, when designing ideals for ecologically sustainable agriculture in any specified country, there is no way of circumventing issues such as transnational links, national agricultural policy, the power and influence of agribusiness, technological traditions, structure of land ownership, social stratification in the countryside, and class- and gender-based divisions of labor.[34] Universal blueprints lose their credibility; instead, more specific solutions are developed with explicit attention to social power relations. In the following, I bring such tensions into sharper focus using "environmental security" as an example.

Environment, Security, and the Military

It seems natural to postulate a connection between security and the quality of the environment: the accumulation of environmental problems increases the vulnerability of nations. Johan Galtung, for instance, emphasized this connection in the early 1980s.[35] Several authors took up the theme by the early 1990s, arguing that problems such as shortages of land, water, or raw materials and transboundary pollution might create new international tensions with the potential for violent eruptions. A comprehensive conception of security thus ought to include mechanisms for solving conflicts arising from such tensions.[36]

However, the concept of "security" requires closer examination because of two interconnected problems. First, it is difficult to define "security" except in negative terms, as an absence of threat. But then it must be specified who (what) is threatened, and by which threat(s). The subject of threat is almost invariably the state. From this follows, second, that any measures strengthening the stability of the state can be considered security-enhancing: security is guaranteed and enforced by the power of the state and its military organizations.[37] However, the security of the nation-state by no means equals the security of the citizens. While preparing for an outside enemy, the state can create direct or indirect threats against the security of the citizens by, for instance, restricting civil liberties and violating human rights, or building up structural obstacles to social and economic equality.[38]

A further problem with attaching the attribute "environmental" to state-based security is that the military itself is not neutral with respect to the environment. On the contrary, environmental warfare, that is, manipulation of the environment for military purposes, has been common in history.[39] This is forbidden by international agreements, but, of course, wars and military operations always cause degradation of the environment, intentionally or unintentionally. For instance, it was well known in advance that the Persian Gulf War in 1991 would destroy the environment in large areas, but the operations were launched nevertheless. The environment does not matter when more traditional security risks are perceived as primary.

Military activities threaten the environment in peacetime also. Armies pollute air, water, and land in their exercises and use natural resources and energy. They are "normal" polluters, comparable to any other large-scale social actors, but they are also "special" polluters that produce particularly toxic and radioactive wastes, as well as "protected" polluters because environmental control of military activities is slack. Thus, the military, especially following the massive development of new arms technologies, including nuclear weapons, is a factor in increasing environmental degradation and environmental risks.[40]

Representatives of the military and defense departments of various countries (e.g., the United States, Germany, Norway, Finland) have recently taken up Galtung's early vision of "environmental security" in discussions on possible new tasks of armies in connection with environmental catastrophes.[41] The argument is that the military as a centralized organization can act efficiently enough to intervene in unexpected situations created by large-scale environmental accidents. Environmental catastrophes are thus presented as "anonymous" enemies that need not be specified but can give apparent legitimacy to the right of the military to use uncontrolled power, even irrespective of national and international law. Examples of tasks that purportedly could be handled by the military include preparation for various emergencies such as oil spills, environmental monitoring using airplanes and submarines, and environmental restoration and cleanup.

This perspective on "environmental security" would thus increase the role and legitimacy of the military. This brings into focus the contradictions inherent in the notion itself. For the traditional type of security to apply to environmental issues, security from violence and security from environmental threats should be similar. However, this is not the case at all. The types of threat posed by war and pollution are very different, their scope is different, and the degree of intention behind the causal agents differs greatly. Most important, organizations that can provide protection from external violence versus environmental problems differ greatly in character: the former are hierarchical, closed, and operate on the basis of secrecy whereas the latter require citizen participation, openness, and free exchange of information.[42] It can also be argued that the link from environmental degradation and resource depletion to international aggression is considerably weaker than implied by the pessimistic scenarios of environmental warfare.[43]

Thus, the concept of "environmental security" appears much less fruitful for environmental protection than some of its proponents have realized. The question is, Would the adoption of the concept actually mean that military logic comes to control ideas about environmental problems? This would greatly strengthen authoritarian tendencies in environmental policy.[44]

Ultimately, the issue in question is the character of ecological problems. Are they sudden emergencies that can be most efficiently dealt with by centralized organizations and authoritarian principles? Or are they, rather, creeping and cumulative effects resulting from a variety of human activities? I think the latter is the case. Enhancing the role of the military in environmental policy leans on a logic that authoritarian solutions are always required, whatever the nature of the problems confronted. This is synonymous with the principle "social order first."

Toward Nondisciplining Political Ecology

I am now ready to weave together the themes discussed in this essay. I argued at the outset that it is impossible to define ecological problems without implicitly including in the definition some idea of social order, that is, of conditions under which the society is itself able to act, or at least to provide possibilities for individual and collective agents to act toward solving the problems. This is the principle "social order first." My review of the views on ecological problems represented by major variants of environmentalism showed that they are saturated by assumptions about the nature of society—

I distinguished from each other "naturalization," "linearization," and "moralization" as main tendencies. Each one of these includes a disciplinary edge, that is, a perception that the maintenance of social order is actually a primary goal in environmental policy, which ordinary people have to serve. The notion of "environmental security" is a particularly straightforward example of such disciplinary logic. It needs to be added, however, that the tendencies are quite heterogeneous in this regard.

Political ecology, understood as an effort to unfold the social and political implications of the ecological crisis, should not fail to take into account the logic "social order first." The question is, What kind of order? The existing social order is usually accepted as the starting point. The record of the existing order should be taken into account, however. The record is quite poor in two respects. First, there is bias, or even sheer neglect, in how environmental problems are defined. Second, the existing order actually contains within itself positive feedback loops that tend to sharpen environmental problems instead of alleviating them. Allan Schnaiberg has called this the "treadmill of production."[45] The problem with general schemes, such as that of Schnaiberg, is that they become unsurpassable. The question is how to change the feedback, or create a favorable, rather than a harmful, one.[46]

My thesis is that solidarity is a key concept in this connection. Before elaborating on this, however, we need to ask what structures and mechanisms can possibly be identified that maintain and enforce the principle "social order first." An important background of this principle is the culture-nature dualism. The main mechanism is objectivization, which implies the subjection of nature and the environment to a master logic.[47] It is, however, doubtful whether an "alternative" can be presented on this totalizing level because a total alternative becomes imprisoned by the present, albeit as its mirror image, and a new dualism is created, whereas dualisms should be discarded altogether. Thus, the first imperative is to avoid totalizing dualisms. What is needed for further specifying the demand is first to identify the opposites in the prevailing dualism and then to find new historically and contextually sensitive ways of combining them.

What, then, are the major elements in the dualism that supports the prevailing logic "social order first" in thinking about the environment? I think the main poles of this dualism can be crystallized as follows: On one side we have the modern society equipped with governmental institutions that are supposed to guarantee its rational functioning. The sciences provide tools for detecting problems and finding solutions to them. On the opposite side is "nature" viewed as an external environment that poses problems, and people who tend to increase such problems with their inadequate conduct.

This dualism needs to be deconstructed. What I called "the ambiguity of Chernobyl" demonstrates that "nature" is not external but also is inside the human body and society. Thus, the culture-nature relationship needs to be conceptualized in a way different from the straightforward externalization of nature. A useful general approach here is drawing a distinction between "first nature," that is, the natural given, and "second nature," that is, the given created by previous human activities. William Cronon's history of Chicago is a good example of the potential of this pair of concepts for specific analyses.[48] The answer on this level is simply to state the fact that culture and nature are mixed together; they are not opposite poles in a dichotomy but merge into complex systems of interactions.[49]

A major issue in integrating human-induced processes and naturally occurring processes together is adequate scaling. This means viewing ecological problems in their historically concrete shape, and breaking the problems into pieces by developing context-dependent approaches toward practical solutions for specified problems. I used "sustainable agriculture" as an example of a higher-level societal goal that can be successfully addressed only through scale-specific recommendations that take into account idiosyncratic aspects of local history and society. Human actors must be explicitly recognized as integral parts in all culture-nature complexes.[50]

But the nature of the social order also needs to be challenged directly. Simply taking the prevailing order as a given lends credibility to existing authorities. The need is for new ways to combine scientific and technological rationality, on the one hand, and an adequate view of human actors and decision makers, on the other hand.

It is not possible to reject scientific rationality, for two main reasons. First, science is "the only game in town powerful enough to offer a window to a viable common future."[51] Second, science is the only means to perceive such facts as need to be perceived—for instance, changes in the composition of the atmosphere. Such phenomena are not directly observable to any human individual using ordinary sensory capacities. However, this does not mean that the constitution of "global problems," let alone derivation of guidelines for action, is achievable by science alone.[52] But neither is there any need to avoid using science. When it is positioned in specified social contexts, science is a perfectly adequate tool for explicating and evaluating rational alternatives available for human actors.[53]

Solidarity is the decisive requirement for constructing a new view on the relationship between abstract rationality and human action. Solidarity can grow from the acknowledgment of the necessarily interactive nature of the formation of human subjects, whether individual or social. Thus, solidarity has two main directions: sensitivity to problems experienced by others, and sensitivity to factors that create and maintain vulnerability of others—societies, communities, and people in the face of environmental, or environmentally mediated, social disasters. Thus, the goal of solidarity is to increase the ability of people to cope with unexpected problems. This contrasts with giving prescriptive advice from a distance.[54]

There is no solution to ecological problems once and for all. Elaborating a new combination of rationality and solidarity is a practical task. In terms of political theory, this relates to the theme of deliberative politics, which aims at overcoming dualisms within theoretical views about democracy.[55] We have no guarantees that acceptable solutions would already be at hand; but, on the other hand, we have procedures and norms on which to build. Respect for those natural processes human culture depends on is one cornerstone; respect for other people with different practices and experiences is another.

The combination of rationality and solidarity is realized in different ways on the global versus the local scale. Decisive is what kind of processes are started, directions taken, and specific means adopted for reaching particular goals. On the global scale, the discourse is constituted by science, but this does not give to science the ability and legitimacy to design scenarios of "planetary management." Such general scenarios cannot capture the real dynamics of regional and local social processes that determine the fate of the global environment, and they cannot but fall prey to the disciplinary logic "social order first."

Notes

1. G. M. Kozubov and A. I. Taskajev, eds., *Radiatsionnoje vozdejstvije na hvojnye lesa v rajone avarii na Chernobyljskoj AES* (Syktyvkar: Institut Biologii, Komi Nauchnyj Tsentr, 1990); G. M. Kozubov and A. I. Taskajev, *Radiobiologicheskie I radioekologicheskie issledovanie drevesnih rastenii* (Saint Petersburg: Nauka, 1994).
2. Based on a paper presented by P. Mardna at the fifth Estonian ecology conference, 1991.
3. Roughly a quarter of the eighteen hundred nuclear tests in the period 1945 to 1989 took place in the atmosphere; see Michael Renner, "Assessing the Military's War on the Environment," in *State of the World 1991* (New York: Norton, 1991), 132–52. For the U.S. experience, see Mike Davis, "Dead West: Ecocide in Marlboro Country," *New Left Review* 200 (1993): 49–73.
4. Yrjö Haila and Richard Levins, *Humanity and Nature: Ecology, Science and Society* (London: Pluto Press, 1992).
5. This is one dimension of Ulrich Beck's concept of "risk society"; see Ulrich Beck, *Risikogesellschaft: Auf dem Weg in eine andere Moderne* (Frankfurt am Main: Suhrkamp Verlag, 1986).

6. This is a central theme in the work of René Girard; see his *Violence and the Sacred* (Baltimore: Johns Hopkins University Press, 1977).
7. Michel Foucault, "Governmentality," in *The Foucault Effect: Studies in Governmentality with Two Lectures by and an Interview with Michel Foucault*, ed. Graham Burchell, Colin Gordon, and Peter Miller (Chicago: University of Chicago Press, 1991), 87–104.
8. Ibid., 102.
9. Zygmunt Bauman, *Legislators and Interpreters* (Cambridge: Polity Press, 1987).
10. Gerard L. Young, "Environment: The Term and Concept in the Social Sciences," *Social Science Information* 25 (1986): 18–124.
11. Beck, *Risikogesellschaft*.
12. Henri Lefebvre, *The Production of Space* (Oxford: Blackwell, 1991); Michel Foucault, *Power, Knowledge: Selected Interviews and Other Writings, 1972–1977* (New York: Pantheon Books, 1980).
13. Such fluctuations in public risk perception are also reflected in the success of Green parties in elections; see Dick Richardson and Chris Rootes, eds., *The Green Challenge: The Development of Green Parties in Europe* (London: Routledge, 1995).
14. For implications concerning research methods, see Allan Stewart-Oaten, William H. Murdoch, and Keith R. Parker, "Environmental Impact Assessment: 'Pseudoreplication' in Time?" *Ecology* 67 (1986): 929–40. Concerning scientific expertise in controversies, see Steven Yearley, "Skills, Deals and Impartiality: The Sale of Environmental Consultancy Skills and Public Perceptions of Scientific Neutrality," *Social Studies of Science* 22 (1992): 435–53.
15. The problem of proving an environmental threat is analogous to the problem of proving the gas chambers. "To have 'really seen with his own eyes' a gas chamber would be the condition which gives one the authority to say that it exists and to persuade the unbeliever. Yet it is still necessary to prove that the gas chamber was used to kill at the time it was seen. The only acceptable proof that it was used to kill is that one died from it. But if one is dead, one cannot testify that it is on account of the gas chamber" (Jean-François Lyotard, *The Differend: Phrases in Dispute* [Manchester: Manchester University Press, 1988], 3).
16. It is quite striking how, for instance, the general environmental concern often neglects the deterioration of inner-city neighborhoods as if the latter were not an instance of the former. There are black spots and white, uncharted territories within the supposedly common concern for the environment; this is analyzed, for instance, in Andrew Ross, *The Chicago Gangster Theory of Life: Nature's Debt to Society* (London: Verso, 1994).
17. The new concerns were officially recognized in the 1972 UN conference for the Environment in Stockholm. Arguments linking traditional conservation and human population growth are summarized by, for instance, Paul Ehrlich and Anne Ehrlich, *Extinction: The Causes and Consequences of the Disappearance of Species* (New York: Random House, 1981).
18. Donald Worster, *Nature's Economy: A History of Ecological Ideas* (Cambridge: Cambridge University Press, 1985), 105–6.
19. Paul Ehrlich's "Eco-Catastrophe!" was originally published in *Ramparts* (September 1969), and has been included in several subsequent anthologies.
20. Geometric growth follows from the assumption that the *growth rate per capita* is constant.
21. Garrett Hardin, "Living in a Lifeboat," *BioScience* 24 (1974): 561–68.
22. On Malthusianism, see Raymond Williams, *Problems in Materialism and Culture* (London: Verso, 1980); Robert M. Young, *Darwin's Metaphor: Nature's Place in Victorian Culture* (Cambridge: Cambridge University Press, 1985); Peter Taylor and Raúl

García-Barrios, "The Dynamics of Socio-Environmental Change and the Limits of Neo-Malthusian Environmentalism," in *The Limits to Markets: Equity and the Global Environment*, ed. Timothy Mount, Henry Shue, and Mohamed Dore (Oxford: Blackwell, 1996).

23. Mike Davis, *City of Quartz* (London: Verso, 1990), 111.
24. This is not only true among biologists. As discussed by William Milberg, "Natural Order and Postmodernism in Economic Thought," *Social Research* 60 (1993), 255–77, naturalization of social order has been endemic in economic theory since Adam Smith, which, of course, makes it all the more difficult to get rid of it.
25. Allan Schnaiberg, *The Environment from Surplus to Scarcity* (Oxford: Oxford University Press, 1980); Gwyn Prins, "Politics and the Environment," *International Affairs* 66 (1990): 711–30.
26. Arthur F. McEnvoy, "Toward an Interactive Theory of Nature and Culture: Ecology, Production, and Cognition in the Californian Fishing Industry," in *The Ends of the Earth: Perspectives on Modern Environmental History*, ed. Donald Worster (Cambridge: Cambridge University Press, 1988), 211–29.
27. Peter J. Taylor, "Re/constructing Socioecologies: System Dynamics Modeling of Nomadic Pastoralists in Sub-Saharan Africa," in *The Right Tools for the Job: At Work in Twentieth-Century Life Sciences*, ed. Adele E. Clarke and Joan H. Fujimura (Princeton, N.J.: Princeton University Press, 1992), 115–48. The ideological background is elaborated in Peter J. Taylor, "Technocratic Optimism, H. T. Odum, and the Partial Transformation of Ecological Metaphor after World War II," *Journal of the History of Biology* 21 (1988): 213–44.
28. The quotation is by Robert Kates, cited in William C. Clark, "The Human Ecology of Global Change," *International Social Science Journal* 121 (1989): 315–45, 317.
29. See also Peter J. Taylor and Frederick H. Buttel, "How Do We Know We Have Global Environmental Problems? Science and the Globalization of Environmental Discourse," *Geoforum* 23 (1992): 405–16.
30. Ulrich Beck, "From Industrial Society to Risk Society: Questions of Survival, Social Structure and Ecological Enlightenment," in *Cultural Theory and Cultural Change*, ed. Mike Featherstone (London: Sage, 1992), 97–124; André Gorz, "Political Ecology: Expertocracy versus Self-Limitation," *New Left Review* 202 (1993): 55–67.
31. This was a dominant theme at the UN "Earth Summit" in Rio de Janeiro in June 1992; Tom Athanasiou, "After the Summit," *Socialist Review* 22, no. 4 (1992): 57–92.
32. Amazingly, this obvious problem is seldom analyzed, but see Marcy Darnovsky, "Stories Less Told, Histories of US Environmentalism," *Socialist Review* 22, no. 4 (1992): 11–56.
33. Carolyn Merchant, *Radical Ecology: The Search for a Livable World* (London: Routledge, 1992), has useful, although quite uncritical, material on this point.
34. These problems are, furthermore, scale-specific such that different issues are most important depending on whether the focus is on national policies, regional production goals, or economy of single farms. See Haila and Levins, *Humanity and Nature*, 167–81, and E. C. Lefroy, J. Salerian, and Richard J. Hobbs, "Integrating Economic and Ecological Considerations: A Theoretical Framework," in *Reintegrating Fragmented Landscapes: Towards Sustainable Production and Nature Conservation*, ed. Richard J. Hobbs and Denis A. Saunders (New York: Springer Verlag, 1993), 209–44.
35. Johan Galtung, *Environment, Development, and Military Activity: Towards Alternative Security Doctrines* (Oslo: Norwegian University Press, 1982).
36. Peter H. Gleick, "Environment and Security: The Clear Connections," *Bulletin of Atomic Scientists* 41 (April 1991): 17–21.
37. Barry Buzan, *People, States and Fear: An Agenda for International Security Studies in the Post-Cold War Era*, 2d ed. (Boulder, Colo.: Lynne Rienner, 1991); Simon Dalby,

"Security, Modernity, Ecology: The Dilemmas of Post-Cold War Security Discourse," *Alternatives* 17 (1992): 95–134.

38. Buzan, *People, States and Fear*.
39. Arthur Westing, "The Military Sector vis-à-vis the Environment," *Journal of Peace Research* 25 (1988): 257–64.
40. Michael Renner, "Assessing the Military's War on the Environment," in *The State of the World 1991* (Washington, D.C.: Worldwatch, 1991), 132–52. The issue is particularly urgent in ecologically fragile but militarily heavily covered areas such as the arctic; see Lassi Heininen, "The Military and the Environment: An Arctic Case," in *Green Security or Militarized Environment*, ed. Jyrki Käkönen (Alderschott, U.K.: Dartmouth, 1994), 153–65.
41. The theme was emphasized, for instance, by the minister of foreign affairs of the Federal Republic of Germany, Hans-Dietrich Genscher, at the opening of the follow-up meeting of the European Security Conference, Helsinki, March 24, 1992.
42. Daniel Deudney, "Environment and Security: Muddled Thinking," *Bulletin of the Atomic Scientists* 41 (April 1991): 22–28.
43. Daniel Deudney, "The Mirage of Eco-War: The Weak Relationship among Global Environmental Change, National Security and Interstate Violence," in *Global Environmental Change and International Relations*, ed. Ian H. Rowlands and Malory Greene (London: Macmillan, 1992), 169–91.
44. Matthias Finger, "The Military, the Nation State and the Environment," *Ecologist* 21 (1991): 220–25.
45. Schnaiberg, *The Environment from Surplus to Scarcity*; Allan Schnaiberg and Kenneth Alan Gould, *Environment and Society: The Enduring Conflict* (New York: St. Martin's Press, 1994).
46. Alan Carter, "Towards a Green Political Theory," in *The Politics of Nature: Explorations in Green Political Theory*, ed. Andrew Dobson and Paul Lucardie (London: Routledge, 1993), 39–62.
47. Feminist scholars, in particular, have emphasized this point; see, for instance, Van Plumwood, *Feminism and the Mastery of Nature* (London: Routledge, 1993). In some of the works there is, however, a tendency to regard the dualistic master logic as a comprehensive and unsurpassable structure.
48. William Cronon, *Nature's Metropolis: Chicago and the Great West* (New York: Norton, 1991). The distinction is perceptively discussed on a more general level by Neil Smith, *Uneven Development: Nature, Capital and the Production of Space*, 2d ed. (Oxford: Blackwell, 1990).
49. Richard Levins and Richard Lewontin, *The Dialectical Biologist* (Cambridge: Harvard University Press, 1985); Bruno Latour, *We Have Never Been Modern* (Cambridge: Harvard University Press, 1993).
50. This point is elaborated by Peter J. Taylor in "Re/constructing Socio-Ecologies."
51. C. Dyke, "Revolution in Changing Environments: Beyond the Balancing Act Frame," forthcoming in *Natural Contradictions: Perspectives on Ecology and Change*, ed. Yrjö Haila and Peter J. Taylor.
52. Taylor and Buttel, "How Do We Know We Have Global Environmental Problems?"
53. Levins and Lewontin, *The Dialectical Biologist*; Haila and Levins, *Humanity and Nature*; Donna Haraway, "Situated Knowledges: The Science Question in Feminism and the Privilege of Partial Perspective," in *Simians, Cyborgs and Women: The Reinvention of Nature* (New York: Routledge, 1991). Ulrich Beck's idea that the modern "risk society" is a laboratory where various societal schemes are put into practice before the consequences are known is also relevent in this context; see his *Gegengifte: Die organisierte Verantwortlichkeit* (Frankfurt am Main: Suhrkamp Verlag, 1988).

54. For philosophical reflections on this point, see, for instance, John O'Neill, *Ecology, Policy and Politics* (London: Routledge, 1993). The question of vulnerability is discussed by Michael Watts and Hans G. Bohle, "The Space of Vulnerability: The Causal Structure of Hunger and Famine," *Progress in Human Geography* 17 (1993): 43–67; and Piers Blaikie, Terry Cannon, Ian Davis, and Ben Wisner, *At Risk: Natural Hazards, People's Vulnerability, and Disasters* (London: Routledge, 1994).
55. Jürgen Habermas, "Three Models of Democracy," *Constellations* 1 (1994): 1-10; Seyla Benhabib, "Democracy and Difference: Reflections on the Metapolitics of Lyotard and Derrida," *Journal of Political Philosophy* 2 (1994): 1–23.

*Over*population the World: Notes toward a Discursive Reading[1]

Saul E. Halfon

United States international population policy is billed as a solution to the "global population problem." More accurately, it reflects concern with the problem of *over*population, especially in the Third World. The "over" in "overpopulation" carries with it a sense of urgency, failure, and fear. It indicates that there are, or soon will be, too many people to achieve our ideal social world—our social imaginary. Such ideals, however, are always normative, and always political. What sort of world is this ideal? What are its parameters? Who is included and who is not? Who will ultimately benefit and who will lose out? These questions are both raised and hidden by a focus on overpopulation, suggesting that this term itself may designate a powerful conceptual apparatus for ordering and shaping the world. A focus on the rise and effects of overpopulation, then, involves more than a study of global population trends and their consequences; it necessitates a political reading of overpopulation as a system of power.

A focus on two intertwined concerns will produce this reading. First, why is there an emphasis on overpopulation in U.S. international policy? Although this question is often answered by a simple appeal to either science or ideology, the science of demography is too contradictory to provide a definitive conclusion and ideological interpretations inadequately deal with the contradictions, motivations, and effects of population policy. Focusing instead on overpopulation as a discourse, which contingently links together many rhetorics, practices, institutions, objects, and categories, provides a better sense for the complexities of this issue and provides avenues for imagining interventions.[2] Second, how is overpopulation discourse implicated in the production and maintenance of domestic politics and identity—what does this discourse achieve? This refocuses attention from the object of discourse to the discourse itself, thus avoiding the usual questions about effective versus noneffective or coercive versus noncoercive population policies. Although at least the latter question is important, both tend to overlook the hidden power relations that are immanent in the production of overpopulation discourse. Focus is instead maintained on how overpopulation discourse establishes powerful fields for surveillance, intervention, and the construction of domestic political and social identity while naturalizing and sustaining discourses around sex and bodies, economics and security, sameness and difference.

This essay tells and retells various elements of the history of U.S. international population policy since the Second World War. First providing a broad political history of population policy, it then links this history successively to changing ideas about the relation of overpopulation to overseas development, especially as that knowledge is produced within the field of demography; the relation between overpopulation and environmental discourse, particularly as that linkage informs public rhetoric; and the discursive construction of various "bodies" within the history of overpopulation discourse. In following this iterative path, it becomes clear that overpopulation discourse is not singular or stable, but rather draws on numerous related discourses in subtle, changing, and sometime contradictory ways. While shifting, overpopulation discourse is nevertheless powerfully enduring within national and international politics. This essay, therefore, maps the domain of overpopulation within the discursive terrain of which it is a part—including institutional, disciplinary, political, and material elements of this discourse. It is in this discursive realm that knowledge and power come together to create and sustain each other.[3]

U.S. and International Population Policy

From its inception in the late 1950s to its form under the Clinton administration, the terms, methods, and rationales for U.S. population policy[4] have varied widely, incorporating many different national and international concerns.[5] U.S. population policy can be broken down into four stages. In the 1940s and 1950s, population concerns were not explicitly emphasized in policy, but were rather incorporated within broader development initiatives. From the 1960s through the 1970s, the linking of development and security concerns with a crisis framing of ongoing Third World famines produced the "population control" policy framework. During the Reagan and Bush presidencies, policy framing was produced through domestic conflicts over development and abortion, resulting in a de-emphasis on population policies. Finally, under President Clinton, action against overpopulation was reemphasized within the "women's empowerment" framework, with a broadened appeal to global and regional environmental and women's rights concerns.

Former U.S. ambassador Marshall Green begins his chronology of U.S. population policy with President Johnson's 1965 State of the Union address, in which Johnson discussed the "explosion in world population."[6] This first presidential statement in support of an explicit population program, however, merely signaled the end point of a gradual shift in U.S. population policy. In the preceding decades, population stabilization had been conceptualized as merely a positive outcome from development programs more generally, under the emerging "population control" framework. Under Johnson, the explicit and targeted reductions of overseas populations came to be seen as both crucial and feasible.[7] This shift was produced through many internal government and foundation studies and documents that reframed the population issue as an imminent crisis that was central to Cold War concerns. Key among such federal documents was a 1958–59 report to the postwar Committee on the U.S. Military Assistance Program entitled "The Population Explosion," which borrowed extensively from Ansley Coale and Edgar Hoover's groundbreaking 1958 study on the population/development linkage in India.[8] Both the 1958 study and the 1958–59 report highlighted population growth's unfavorable effects on economic growth, which at that time was seen to have negative consequences for the social and political stability of the Third World, a concern that was heightened by the perceived threat of Communist expansion.[9] Such "behind-the-scenes" work on overpopulation helped to produce a body of thought and a constituency of people in the executive branch, congress, private organizations, and public groups that slowly articulated a need to more directly integrate population concerns into the country's development program.[10]

As the linkages between these various groups and their agendas grew, strengthened, and became institutionalized within private organizations, USAID, and the growing field of demography, the United States began to take a leading role in the production of an internationalized discourse on overpopulation. In particular, by extending financial and technical assistance, these groups became foundational for the implementation of population control strategies throughout the world. Finkle and Crane describe how the robustness of the linkages between these various groups and organizations allowed the population control establishment to carry its agenda through the 1974 International Population Conference in Bucharest, despite strong opposition from many Third World countries.[11] Many of these countries viewed the U.S. position as an ideological, neocolonial attack on their sovereignty, and used the conference as an opportunity to push instead for increased international economic equity—what the "Group of 77" countries called the "New International Economic Order." Although a consensus "plan of action" was produced from this conference, which did address overpopulation as a potential problem, this plan did not contain the strong "control" language that was by then commonplace within U.S. policy circles. It framed overpopulation instead as a symptom of economic inequality and a regionally specific phenomenon. Nevertheless, this event had very little impact on the U.S. position through the rest of the decade. USAID, the World Bank, the Population Council, and other organizations continued to implement the U.S. population control agenda through numerous targeted, locally based programs.

Such population control policies had strongly emphasized technical improvements in birth-control technology and distribution as the best way to deal with rapid population growth. These policies assumed that women would control their births if only they had the contraceptive resources to do so. During the 1980s and into the 1990s, this notion was challenged on a number of fronts, particularly by feminists, anthropologists, and critics from the Third World. Such critiques challenged basic assumptions about women's motivations and actions, questioned the importance of modern contraceptives for achieving lower birthrates, exposed unfounded assumptions about various cultures, and leveled ethical, moral, and political attacks on the effects of earlier policies.

These critiques, however, were largely downplayed by the population establishment throughout the 1980s, when the U.S. policy coalition was disrupted by another set of concerns. For the first time since the early 1960s, the assumptions of the population establishment were fundamentally challenged at the highest levels of government. Pushed by both antiabortion activists and a new coalition of conservative, free-market economists, the Reagan administration rejected the "population control" framework. During the early 1980s, the Reagan administration cut or withheld funding for the United Nations Population Fund, in response to UN support for the population control programs in China, many of which were seen to be coercive. They also eliminated all funds for the International Planned Parenthood Federation because of its worldwide support of abortion. Although these two actions sparked tremendous controversy and prompted numerous congressional hearings and bills, it was, again, an international population conference that consolidated the "official" U.S. position.[12]

In preparation for the 1984 World Population Conference in Mexico City, the State Department (under the antiabortion delegate to the conference, James Buckley) drafted its first Reagan-era policy statement on population and development. Inspired largely by Reagan's economic advisers, this statement declared population to be a "neutral phenomenon" with respect to development.[13] There was, therefore, no justification for pushing population programs beyond the humanitarian and democratic goal of simply providing people the resources, other than access to abortion, necessary to attain the family size that they desired. Again, the United States became the outsider at this conference. This time, many Third World countries were themselves developing population control programs and seeking international resources and support for these activities. Finkle and Crane attribute this turnaround in the Third World to increasing economic depression worldwide and the belief that increasing populations were exacerbating this situation.[14] It is also important to note, however, that most of these programs were being created by the elites of these various countries, who were continuing to receive both scientific and financial aid from elements of the pre-Reagan population control coalition.[15] The ability of these elites to position themselves in opposition to "U.S. hegemony" provided further justification and legitimation for such control programs.

Although President Bush loosened some of the funding restrictions imposed during the Reagan administration, the hands-off policy continued until President Clinton was elected to office. On May 11, 1993, the Clinton administration sharply diverged from the U.S. population policy of the previous decade. In his May 11 statement, Timothy E. Wirth, the undersecretary of state for global affairs, announced that the administration would resume contributions to the UN Fund for Population at early 1980s levels, and would make population a central part of development policy in general. This was reinforced on January 11, 1994, when the administration declared that limiting population growth was its "top priority" in foreign aid.[16]

Although reminiscent of pre-Reagan/Bush population policies, the Clinton policy differed in some important ways. Drawing on the many critiques of the population control framework that had been developed over the previous decade, policymakers no longer believed it to be adequate merely to increase access to Contraceptives. Birthrates were seen as being determined by a complex set of factors related to women's desire and ability to control their reproduction. These new ideas have been captured within the growing "women's empowerment" framework.

As with previous policy frameworks, an international population conference served as a crucial forum for the articulation and adoption of the "women's empowerment" framework. At the 1994 Conference on Population and Development in Cairo, the United States once again took the lead—this time with a renewed emphasis on supporting international efforts in family planning. The main difference between this and previous conferences was the strong emphasis on the role and status of women in the Third World, including their importance for Third World development. Overpopulation is thus, once again, becoming strongly tied to development, this time via the empowerment of women through the provision of basic health services and family planning, basic education, job training, and other similar programs. The theories of a direct connection between population and development have been sidelined in favor of intervening institutional and sociocultural mechanisms related to poverty, education, female independence, security, and social customs. Furthermore, the notion of development that is invoked is much less focused on national economic measures; instead, it treats human welfare and choice as laudable development goals in themselves.

This history demonstrates some of the shifting concerns and discourses that have produced and motivated U.S. population policy. The next section will expand this political history to explore the role that demography has played within policy formation. Although changes in demographic theory play a role in producing particular policy frameworks, policies around overpopulation have not merely been justified through a simple invocation of numbers of people and their distribution in the world. Rather, overpopulation policies, and the notion of overpopulation itself, have been constructed through and within a rich matrix of scientific, political, historical, and cultural discourses. Tracing out this matrix will provide insight into what overpopulation is, how it is sustained, and what it achieves.

Modernization and Third World Development

Changing theories of development and ideas about modernization have been deeply intertwined with changes in population policy. Since the linkage between population and development has been largely articulated within the field of demography, tracing the history of demographic theory around this link is important for understanding population policy. Following Dennis Hodgson's periodization, ideas about the population/development link shifted during the postwar era from Demographic Transition Theory through "orthodoxy" and "revisionism" to a form of cultural demography.[17] Although these theoretical frameworks have not fully replaced each other, they do highlight the growing conflicts within the field and the ways that such conflicts link with policy production. Exploring the political contexts of such scientific narratives reveals how these ideas map onto the political and cultural context within which overpopulation has been produced.

Demographic Transition Theory (DTT) holds that populations progress through a series of stages as a society develops.[18] "Traditional" societies are characterized by high fertility and high mortality rates. These populations tend to remain stable or grow very slowly as these two variables balance each other. As a country develops, its mortality rate tends to fall somewhat faster than its fertility rate and the country experiences a rapid increase in population. Eventually, because of the social and economic changes that accompany development, the fertility rate decreases to a level that is equal to or below the mortality rate, thereby producing a stable or shrinking population. In extending this theory from historical studies of European populations to the Third World, a new variable was introduced: colonialism. The transition theorists realized that colonial governments often introduced new public health measures and medical technologies into societies, resulting in rapid decreases in mortality. However, since this was not "true" development, the social, economic, and cultural changes that lead to lower birthrates developed at a much slower rate than in noncolonized countries. Thus, these societies experience unprecedented rapid—and, from a neo-Malthusian perspective, dangerous—population growth.

DTT thus located development itself as the primary issue, with population growth as a secondary manifestation. It placed motivation as the key factor in birth control, with increased development as the best route to motivating lower fertility. Within this perspective, an explicit "population policy" made little sense. Population concerns were instead couched within broader development goals, such as technological and agricultural development.

The relation of this theory to overseas policy can be best understood by reference to the context in which it was produced. DTT was formulated at precisely the time that demography was first being institutionalized as a discipline and that modern "development discourse" was itself being formulated.[19] Thus, the linkages between these various phenomena were articulated quite strongly. Development discourse arose during the 1930s and 1940s within both macroconfigurations around the expansion of markets and the loss of stable materials during the collapse of European empires, and within the micropractices of economistic theories and ideas.[20] The rise of overpopulation as a concern involved the articulation of such modernization ideals with racial and nationalistic ideas, which drew heavily on the eugenic theories of the 1920s and 1930s, but increasingly invoked political fears about the growing power and global significance of Asian countries.[21] DTT thus provides a strong relation between population size and modernization, arising at the intersection of concerns around economics, political stability, racial and ethnic fears, and disciplinary development.

Although DTT is still considered foundational to demographic analysis, in practice it was replaced in the 1960s by what is commonly called "orthodoxy." This new theory took shape in the 1950s and 1960s around a set of challenges to DTT. In particular, many studies started to suggest that the European model did not hold, especially because development was itself being hampered by the growing populations. The ensuing shift away from the traditional DTT framework was reinforced by a new framing of overpopulation as an environmental crisis, coupled with cold-war fears of underdevelopment and the close relations between demography and its activist funding sources.[22] These developments and the production of "orthodoxy" contributed significantly to the rise of the first explicit population policies, within the "population control" framework.

The orthodox position was initially framed in a classic study by Coale and Hoover.[23] Coale and Hoover set out to model the relation between demographic and economic growth in India. In the macroeconomic model that serves as their framework, population growth necessitates that already scarce resources be used to support a growing, child-heavy population rather than being used for capital investment. Unlike more developed areas in which a growing labor pool can be utilized to increase production, these underdeveloped countries do not have a resource and capital surplus, and are thus caught in a zero-sum situation. Although Coale and Hoover agreed that a resource crisis of the sort that Malthus proposed was not imminent, they claimed that the stagnation of economic development that results from demographic growth would eventually lead to such depletion. They saw the best solution, therefore, in quick action toward slowing such population growth, in order to spur underdeveloped economies. The assumptions behind this early study were steadily chipped away in subsequent years, but it nevertheless formed the starting point for many theorists and policymakers sympathetic to Coale and Hoover's ultimate conclusion: that direct reduction in birthrates is necessary for the development of many nonindustrialized countries.[24]

Translated into policy, the orthodox position holds that lowering population must be achieved directly by supplying modern contraceptives to people in the Third World. This position is called the "population control" framework in international policy, which is characterized by a focus on contraceptive delivery and target-driven birth-reduction efforts. Thus, the orthodox position, while still addressing the effect of development on overpopulation, adds the reverse link and raises the importance of overpopulation within overseas policy efforts.

In the late 1970s, a new perspective, driven primarily by free-market economists, began to emerge. This "revisionist" perspective questioned the most basic assumption of earlier periods: "that rapid population growth was a serious problem."[25] In its most extreme form, this view postulated that population growth was both beneficial to and necessary for development because high population density spurs the technological, social, and economic innovations necessary for development.[26] While strengthened by growing evidence of the ineffectiveness of population control programs, revisionism got its biggest political boost within the Reagan administration, which embraced this strongly free-market-oriented theory, and translated it into the official U.S. position at the 1984 international population conference in Mexico City.[27] Revisionist economist Julian Simon became the Reagan administration spokesman on overpopulation.

Although revisionism never became the dominant position, either in demography or population circles (despite Reagan's attempts), it has served to undermine the strong and unquestioned position of population control strategies. It has also played an important role in driving demography and many demographers toward a more neutral position with respect to population size, and has helped them to begin loosening some of the more obvious intellectual links to the policy arena.

A more recent response to orthodoxy moves past the strictly economist view, in ways hinted at by Coale and Hoover themselves, who recognized that production is not based simply on the amount of money and number of workers being utilized, but is also dependent on social institutions that increase the usefulness of these two factors of the production process. Coale and Hoover mentioned the importance of educational institutions for increasing the productivity of workers and, therefore, increasing the useful capacity of the already available labor force.[28] Such educational institutions could therefore shift the effect that a particular population size or growth rate has on the economy or society as a whole. From this starting point, other researchers have begun to explore numerous other social institutions affecting economic growth. In summarizing this line of theorizing, the National Research Council's Working Group on Population Growth and Economic Development stated in 1986 that "these observations point to the key mediating role that human behavior and human institutions play in the relation between population growth and economic processes."[29]

This focus on institutions and other sociocultural factors has given way to views of development that are not wholly reliant on an economic perspective. Education, health services, women's empowerment, changes in family structure, democratization, and technological innovation have all begun to be seen as laudable ends in themselves and important steps in the process toward modernization. In this view, economic growth is only one element of the development process. Subsequently, even the teleology inherent in the notion of modernization has begun to be questioned.[30] Definitions of development have started to move away from the idea that the modern Western world is the end point for all nations. They focus instead on defining development as a process in which more localized social, economic, and cultural goals can be met.[31] Although such cultural sensitivity has long been expressed in terms of the need for and content of population policies themselves (within U.S. policy circles), such ideas have only recently been incorporated into the very definitions of development that are used in the creation of population strategies.[32]

In the 1990s, this developing perspective facilitated the replacement of the "population control" policy framework with the "women's empowerment" framework. Building on both anthropological and feminist critiques of the orthodox and revisionist positions, this new framework targets women's familial and broader cultural contexts as crucial sites in the production of population behavior.[33] This framework relies on various demographic theories and their critiques. From DTT it borrows the notion that broader changes in society and culture are necessary in order for people to alter their reproductive behavior. From orthodoxy it borrows a focus on both micro-level practices and the need for reproductive services. However, it tends to reject both DTT's and orthodoxy's broad homogenization of the development process and the individuals and societies involved. Finally, from revisionism this framework borrows the vision of overpopulation as a secondary issue, while rejecting the free-market orientation of this perspective. Instead, it holds that women should be provided with locally specific means for gaining control of their own lives because that is the right thing to do, and if population problems are helped along the way, all the better. This policy framework was most clearly articulated within the context of the 1994 International Conference on Population and Development in Cairo.

In short, while the relation between U.S. policy making on overpopulation and demographic theories is strong, policy did not spring solely from these theories. The concern over population sizes overseas arose within a context of economic expansion, political fear, and important shifts in the global political and economic order: demographic theories bound these concerns into a scientific object, replete with objective frameworks for action. Within this scientized context, the population/development link has consistently served as the foundational and least politicized of the various links constituting overpopulation discourse.

The Global Environment

Economic and social development provide much of the framework for policy thought. However, in recent decades environmental concerns have also driven public ideas about the need for population control. Authors such as Paul and Anne Ehrlich, Garrett Hardin, and Lester Brown are the most visible spokespeople for what may be termed neo-Malthusian environmentalism. Although concerns over the environment have long sustained action around population size—even the dilemma detailed by Malthus was essentially about the interaction of expanding populations with their environments—the nature and centrality of such concerns have shifted dramatically over the years. In particular, the shift from a focus on the localized famines of the 1960s to a globalized "crisis" rhetoric in more recent years raised the stakes of overpopulation discourse and provided strong incentives for the policy interventions characterizing the "population control" framework.

In the mid-1960s, a series of famines within African and Asian countries were interpreted by policy analysts as resulting from rapid population growth. Such concerns provided a strong public impetus and justification for targeted population control efforts.[34] These population concerns were not, at the time, seen as strictly "environmental," but rather were linked fundamentally with issues of political stability, economic expansion, and the spread of communism.[35] By the early 1970s, however, the localized focus on specific famines was supplanted by a growing globalization of analysis, including theorizing about natural limits and worldwide sustainability. Books such as *The Population Bomb* and *Limits to Growth* highlighted the effects that population growth would have on "all of us," on the global environment. When the "bomb" explodes, even the industrialized West will be destroyed by its force. These stories about the environment's relation to population rely on aggregation, homogenization, decontextualization, and the naturalization of social systems—the same approaches that make both the need for and the implementation of a global population policy feasible. In various forms, this vision of the world still predominates among those who militate for population control. "In fact, in the absence of any analysis of differentiated interests, population discourse logically offers no other standpoints for an environmentalist to take."[36]

Although such neo-Malthusian positions have driven the overpopulation crisis in recent years, numerous critiques have been made of this perspective.[37] Drawing heavily on anthropological and (feminist) ethical and political perspectives, critics have provided much of the groundwork for more recent shifts in theories about overpopulation and population policy. Exploring these critiques will serve both to contextualize some of the shifts toward the "women's empowerment" framework and to provide a clearer sense for some of the political implications of the older "population control" framing.

One focus in recent ethnographic work has been detailed investigation of the complex localized processes that create regional environmental crises. For example, in an article that explicitly addresses the relation between environmental destruction and population change, Raúl García-Barrios and Luis García-Barrios explore the effects of population shifts in Oaxaca, Mexico.[38] They explore how "structural adjustments" within Mexico caused the semi-proletarianization of the population of this region, and rapid movement to the cities. As the villages became depopulated, rather than achieving an increase in per capita resources, food resources declined. This resulted from the collapse of the local social institutions that had maintained the elaborate terracing systems in which agricultural products were grown. As these social structures collapsed, so did the terraces, resulting in a rapid depletion of topsoil through erosion. Thus we have a case in which depopulation, coupled with very localized social structures, resulted in soil erosion and food scarcity.

The need for analytical specificity that this account highlights is supported by numerous other studies on environmental destruction within particular regions. Complex interactions of ecological, political, and cultural factors lead either to ecological sustainability or to ecological destruction (neither of which should be considered as some sort of "natural" state).[39] Such studies strongly undermine standard narratives that either naturalize peasant vulnerability and unsustainability or naturalize and glorify sustainable peasant knowledge, while erasing the historical or localized processes that create such situations.[40] Ignoring such specificity, and thus naturalizing and globalizing the "local" processes through which overpopulation creates environmental destruction and famine, is a political activity. This highlights the political and constructed nature of statements such as "rapid population increases undermine peasants' naturally sustainable practices" or "rapid population increases push already unsustainable systems over the edge." Both of these formulations suggest the need for rapid intervention into population growth itself.

The construction of the globalized dimension of overpopulation has similar implications. Through globalization and the de-differentiation of people in their relation to the environment, numbers themselves *become* problems. Too much or too many translates easily into destruction. The solutions, therefore, take on a mathematical quality: if there are too many, we simply subtract some. This becomes a technocratic imperative—protection of the earth, management of our limited resources, control *for* the future. It presents Earth as a delicate communal system that must be saved at all costs. This imperative cries for action, *urgent* action. It is enframed for us by notions of science, development, and control.[41] The pictures that are painted for us eschew alternative visions of our world. Thus, global discourse can "appeal to *common, undifferentiated* interests as a corrective to scientifically ignorant or corrupt, self-serving or naive governance."[42] Through this translation from method, to fact, to action, a scientific framing of population necessitates a technocratic response.[43]

Nancy Peluso also explores international calls for action.[44] Western environmentalists recognize that global environmental problems can only be resolved in concert with other nations, through a transnational cooperative politics. Thus, environmental treaties give international weight to domestic conservation policies. But these national policies may be used not just to manage nature, but to manage people. Management can become a euphemism for violence, the coercion of reluctant groups, and the militarization of ecopolitics. Ann Anagnost also explores control and management in China's one-child population policies.[45] She indicates that China's policies may be deeply implicated in reproducing the very patriarchal and abusive family structures that are officially denounced by the government. Women in China are therefore exposed to increased violence both in the home and by the state in response to their reproductive actions. According to Anagnost, the formulation of these policies both draws on and is in response to an international definition of modernization that is drawn primarily from the West and is based within notions of control. These two examples clarify Soheir Morsy's warning: "for the objects of the rescue mission themselves, 'Western' compassion can be nothing less than the kiss of death."[46]

Bodies

Although discourses on modernization and the environment have provided a great deal of the epistemological, institutional, and rhetorical motivation for population policy, a look beyond these spheres is necessary for an understanding of the cultural space within which overpopulation discourse functions. How and why do the multitude of concerns and issues discussed earlier get worked out within the context of Third World population control strategies? Recognition of the centrality of bodies within overpopulation discourse provides insight into this question. Whether population policies coerce or persuade or merely provide the resources with which people can make their own reproductive decisions, attempts to affect the rate of population growth converge on reproductive behavior—on changing the social, cultural, and personal configurations around bodies. Importantly, the bodies of concern are rarely those of the primarily white, primarily male, primarily upper- and middle-class (elite) policymakers in the United States; they are rather the bodies of women, of Third World peoples, and of the poor. Since both demographic knowledge and national and international political struggle around overpopulation rely on knowledge about bodies and behavior, overpopulation intersects with broader discourses of gender, race, class, nationalism, ethnicity, sexuality, family, disease, and reproduction.[47] An exploration of these various discourses necessitates asking why particular body discourses appear so prominently and others are never present; why particular articulations are more successful than others; who and what gets produced/reproduced, and who and what gets erased in the production of international overpopulation policy. Three types of body constructions are particularly important in their articulation with overpopulation: gendered bodies, racial bodies, and medical bodies.

Gendered (Female) Bodies

The construction of the gendered body in overpopulation discourse is not simply about men and women; it is about women as reproducers and men as always absent producers/exploiters. This becomes particularly evident within the women's empowerment framework, which focuses on women as the central site of aid, intervention, and empowerment. As Hartmann suggests, "gender has become a way for population agencies to bypass politically sensitive issues of race, class, and inequalities between developed and developing countries. Everyone, it seems, is for empowering poor women these days—but not for empowering poor or marginalized men. But many poor men are also losers in the development process."[48]

It is, then, through the erasure of these other bodies that the gendered (female) body comes to play a central role in overpopulation discourse—the women of concern are always poor, often "dark," usually from the Third World, but this is secondary to their existence as women. The empowerment framework and debates over reproductive rights and abortion provide a vantage point from which to explore the gendered body.

Within both demographic theory and policy production, the framing of women and their reproductive practices has shifted over time. In the earliest formulations, women's bodies appeared primarily as the vehicles through which development practices expressed themselves in birthrates. The various logics of modernization were seen to be interpreted and implemented through women's bodies—whether expressed in the cost calculus of rational choice models, as the users of technological advances, or as the pawns of cultural practices.[49] With the shift to "orthodoxy" and the "population control" framework in the 1960s, women were reinscribed as having unfulfilled reproductive desires, as being autonomous individuals who simply could not act in their rational best interest because of social and technological constraints.[50] Although the perceived importance of development as a way to reformulate the calculus of birth decisions remained, through this period and into the present, aid activity focused more on the provision of contraceptive technologies. Because the "autonomous individuals" often resisted the interests ascribed to them by the implementers of such policies, the deliverers of contraceptives have preferred long-term, passive methods such as IUDs, injections, implants, and sterilization.[51]

From a series of critiques of these past practices, leveled by demographers, feminists, and others, a new construction of women's bodies emerged. Women's reproduction is now seen as cultural: choices are not just economic and rational; rather, they are embedded in a broad array of familial, social, political, and other cultural practices. These practices, which are often repressive and oppressive, are resisted and contested by these women, but often in ways that are deeply mediated, circumscribed, and constrained. "Empowerment" involves providing these women with the resources, knowledge, and desire to resist oppressive structures and to find their own desired levels of fertility. Empowerment is still about "modernizing," but it focuses particularly on targeted modernizations that will enhance women's abilities to effect their own choices, desires, and possibilities.

This shift to empowerment is also produced within another set of discourses around women's bodies. In the West, women's empowerment and reproductive freedom have been strongly tied in the past few decades to conflicts over abortion. Bringing with it a complex package of "rights" talk, linkages between overpopulation and abortion attempt to construct Third World women as "us," as an equivalent group whose futures and lives are tied inextricably to our own, who play out the same conflicts that we face in our lives. The notion of overpopulation, in its connection to abortion and reproductive rights politics, is supported by struggles over political and moral identity in the West.

This connection between the abortion debate and overpopulation has been evident since at least the mid-1980s, when the Reagan administration used concerns over international support for abortion to cut funds to several population groups. This positioning was further solidified in the Reagan administration's firm opposition to population control during the 1984 Mexico City population conference. The Clinton administration alternatively made a strong stand in support of abortion internationally, pitting itself against the Vatican and many Muslim leaders during the 1994 Cairo conference. A closer look at these cases reveals the linkages between abortion and overpopulation working in both directions, with abortion and overpopulation each providing frames and resources for struggles over the other.

The historical conflation of "choice" and "abortion" in domestic abortion debate has had an important impact on how politicians and other political actors can speak about population policies.[52] Despite isolated attempts to keep "choice" and "abortion" apart, as in struggles over the name "pro-choice" versus "proabortion" and "pro-life" versus "antiabortion," this distinction has not been maintained consistently in practice. The effect of this, when combined with the strong political linkage between population control policies and abortion, is to articulate a choice-abortion-population relation that has associated population control policies with a pro-choice stance. General critiques of population policy, therefore, are easily construed within the political sphere to represent an antichoice position, unless the very difficult political task of disarticulating these positions is attempted. Such disarticulations are often achieved by feminist and Third World critics, but they are much rarer within the broader policy arena.

Cultural arguments about the effects of Western reproductive practices, particularly abortion, have also played a significant role in framing positions on overpopulation. For instance, opponents of population policies invoke the possibility of cultural hegemony in the imposition of new technologies and practices on communities. Often, such communities already have numerous practices in place to accomplish similar results, or, if they do not, such results are often counterintuitive from a local perspective.[53] In fact, the new practices may disrupt the social organization of such communities, leaving the very people they are trying to help without the traditional social supports they have come to expect.[54] Alternatively, others argue that new reproductive practices might undermine state structures and cultural institutions that keep people, primarily women, under control, thus expressing emancipatory potential. To these activists, reproductive rights provide a historically grounded and necessary route toward emancipation for women—an essential component of any feminist struggle because it provided much of the groundwork for feminism in the West. They also invoke the symbolic importance of abortion to show that giving women control of their own bodies is an essential feature in allowing them to make the choices that will reduce reproduction, whether or not they actually use abortion as a method to achieve this reduction.

Although abortion has thus informed the overpopulation debate in important ways, the abortion debate is also being fought *through* the debate on overpopulation in ways that broaden the terms of both debates. For some abortion opponents, population policy provides examples to substantiate their fears and a site to extend their critique of abortion rights. For example, the coercive and eugenic potential of international population programs is often clearer than in domestic examples.[55] Likewise, arguments about cultural hegemony can be more easily made about Third World countries because the notion of "cultural difference" is rather easily articulated in such cases. Also, connecting to such an international issue allows abortion opponents to globalize the debate and see themselves as connected to a larger network of concerns, activists, and issues.[56] For proponents of abortion rights, the international aspect of the overpopulation debate provides access to numerous comparative examples of countries with or without substantial reproductive health programs, and allows direct comparisons between conditions in the United States and Third World countries. It also allows activists who are supportive of population reduction to internationalize and scientize the proabortion position by demonstrating that human survival depends on the availability of abortion. This moves debate from personal choice to global and moral imperative. Finally, the focus overseas provides abortion advocates with demonstrative examples of the truly emancipatory potential of reproductive choice—they can point to countries where such emancipation has recently occurred in relation to reproductive rights. In the use of overpopulation as a resource in abortion debates, not only does population become further stabilized as an important issue, but the activists themselves, and their agendas, become increasingly drawn into and tied to larger political, economic, and human rights issues.

This discussion demonstrates ways that abortion and overpopulation have sustained each other as issues, even while the abortion issue has served to confine the positioning available to political actors with respect to population policy. Thus, this link has effectively made it impossible to strategize around reproductive health, abortion rights, and contraception in the Third World without invoking or dealing explicitly with population issues. Although such a link reinforces the concern over both issues, it also shifts the terms of debate in fundamental ways, and provides sites for crucial reformulations of agendas. In fact, overpopulation discourse is often extended, justified, and reinforced in public debate through forms of strategizing that highlight lower population as a positive outcome from women's empowerment.[57] Because these two concerns tend to gain strength in areas in which they are mutually reinforcing, even the feminist positions tend to come out sounding globalized and homogenized within such debates. This, in turn, ultimately reinforces the decontextualization and objectification of women's lives that underpins traditional population policies.

Racial (Dark) Bodies

The discourse of the gendered body within overpopulation articulates similarities between people in the West and the Third World. It constructs the oppressed woman as a "sister," as one of "us" in need. The racial construct, however, produces *dark* bodies as dangerous and premodern, as a place for education, intervention, and control. The dark body is an infantilized "other," inspiring fear.[58]

The racial body is produced through a discourse of "difference" that is often articulated through the body. Thus, the racial body appears throughout the history of overpopulation discourse and population policy. It was evident in postwar fears about the growing power of Asian countries. It was evident in the national security policy that called for rapid population control in key Third World states.[59] It is evident in the ability of Al Gore, vice president under Clinton, to acknowledge the environmental impact produced by the "hugely consumptive way of life so common in the industrial world," and then so easily to dismiss its relative importance with respect to population growth in the Third World: "But the absolute numbers are staggering."[60] All of these rely on an articulation of difference and an expression of fear. In the construction of the racial body resides a long history of eugenics, colonialism, and overseas intervention.

In the late 1960s, the primarily political and military fear that had undergirded international racial discourse in the preceding decades gave way to the reinforcing rhetorics of global environmental threat and Third World bodies out of control. This shift was characterized by Garrett Hardin's "lifeboat ethic," an extension of his notion of the tragedy of the commons, and Paul Ehrlich's *The Population Bomb*, probably the most influential popular work on overpopulation to date. Hardin represents a well-respected extreme in overpopulation debate. "What happens if you share space in a lifeboat?" he asks. "The boat is swamped and everyone drowns. Complete justice, complete catastrophe."[61] In *Science* magazine, he suggested that the "freedom to breed is intolerable," and that coercion is probably the best solution to overpopulation since "injustice is preferable to total ruin."[62] This heightened polemic of crisis, of imminent threat, although downplayed over the next few decades, set the groundwork for the new racial body. It is clear that those swamping the lifeboats and needing a little coercion are not the same people he refers to when he states that "those who are biologically more fit to be the custodians of property and power should legally inherit more."[63] Although Paul Ehrlich suggests a more cooperative solution, with population control in both the "developed" and "undeveloped" countries being important, the emotional context for his discussion resides in his experiences in "an ancient cab . . . one stinking hot night in Delhi."[64] His narrative around this event is presumably about "people, people, people, people," but Mamdani indicates that it is the "quality" rather than the "quantity" of people that gives this story its force and invokes an emotional response—

these people are poor, Third World, terrifying "others."[65] The crisis talk of the "population control" era, then, relies on the production of the racial other. The "dark" bodies in these narratives are undifferentiated—they form masses of people, abstractions, and numbers, who are multiplying and overwhelming. As Chandra Mohanty states: "colonization almost invariably implies a relation of structural domination, and a suppression—often violent—of the heterogeneity of the subject(s) in question. . . . [The] production of the 'Third World Woman' as a singular monolithic subject."[66]

As the racial body moves into the 1990s, it is precisely these elements of undifferentiation that carry through, as the more obvious race talk gets sublimated within a scientized debate over "carrying capacity."[67] Despite feminist and demographic calls for a focus on family and village-level dynamics, and the importance of the World Fertility Surveys in providing detailed, individualistic data, aggregate numbers still predominate in public discussions, and projections of future population sizes reign supreme.[68] Will the world contain eight or fifteen billion people by 2050?—as if this provides any sense for how people will differentially experience the world at that time. And even now, population agencies easily take on the rhetoric of empowering "Third World women"—denying that such women have any agency or discounting the ways they empower themselves—thereby extending the "construction of 'Third World Women' as a homogeneous 'powerless' group often located as implicit *victims* of particular socio-economic systems."[69]

The racial body is also a contested body—contested abroad and at home, often for very different reasons. Abroad, it is contested by those who are racialized, the monolithic subjects themselves. At home, it is contested in the conflicts between the gendering of "sameness" and the racializing of "difference." And it is also contested in the politics of overpopulation itself—through the history of eugenics. Recent invocations of the history of eugenics will provide another glimpse into the modern construction of the racial body.

Few leaders would currently support eugenic policies outright, yet the history of eugenics in America has provided a very important site for emphasizing the implications of population policy.[70] Although this history cannot be read directly onto overpopulation discourse, it does provide a lens for analysis and a set of conceptual resources for various actors. In this light, one does not have to look back very far to note the strong historical connection between birth control and eugenics, or to read that eugenic history back onto more recent population policies.

Eugenics had achieved great currency among the primarily white, middle-class intellectuals and reformers of the 1920s and 1930s. This ideal of race purification grew largely out of concern over increased immigration, African-

American migration to the north, and increasing poverty in American cities. Many of the early birth-control advocates in this country, including Margaret Sanger, emphasized the eugenic notions of slowing the reproduction of immigrants and the poor, whom they feared would soon overwhelm the older European stock or become too large for social services to support.[71] This fear of large populations of "the Other" provides a historical link between population, birth control, and eugenics.

To many population control and some reproductive rights activists, this history merely represents a distant past, a time that serves as a warning but bears no direct relation to the present. For these activists, careful awareness allows practitioners to avoid the historical pitfall of racist and classist eugenics; it allows a deracializing of reproductive bodies in an effort to protect them. To many feminist and Third World activists, current population policies embody a globalization of the same fear that motivated domestic eugenics: a fear of "the Other," of white Americans losing political and economic supremacy, of the Third World depleting a system that was built for "us."[72] This vision produces a diagnosis of the already racialized body, in order to protect it. To still others, primarily those opposed to population control, eugenics serves as a powerful moralistic story to use in their struggle to oppose the extension of population programs and reproductive rights overseas. In this view, easily colonized Third World people are vulnerable to the reassertion of eugenic programs. These activists seek to actively racialize the body, in order to protect it. Such constructions should always be noted within efforts to protect others, if that is actually the goal.

Medical (Sick) Bodies

Another important element in the production of overpopulation discourse is the construction of the medical body. This body is surrounded by systems of treatment, intervention, and surveillance, but also of empowerment and modernization. When gendered (female) or racialized (dark), the medical body often becomes the sick body, reinforcing rhetorics of crisis, need, and danger.

The medical body was re-produced in the shift from the "population control" to the "women's empowerment" frameworks. Within the "population control" framework, women were given contraceptives in clinics dispersed throughout the Third World. These clinics were often separated from health facilities, in an attempt to decrease the opportunity cost for women obtaining contraceptives: nothing was required of them other than their acceptance of the contraceptive itself.[73] However, critics began to call attention to the fact that these centrally and quota-driven clinics were often providing untested products to women, failing to warn them of dangers or side effects, and failing to provide follow-up services. This was not the system of reproductive health delivery provided in the West; it was merely a system for decreasing births. The critics, including many Third World activists and Western feminists, began to call for the reconnection of reproductive services with basic health-care delivery. Community clinics were to become a place in which routine gynecological screening would occur, in which children would receive immunizations, in which nutrition and disease-prevention information could be distributed. They also called for these clinics to become a place in which contraception would be distributed voluntarily, safely, honestly, and with adequate follow-up. The women's empowerment framework, then, reinscribes a medical discourse on Third World women, and serves to medicalize overpopulation.

Such an increased focus on medical intervention has multiple and contradictory implications. Although this framing deals well with the critiques leveled at population control policies, it also inscribes a new logic of modernization. Rather than relying on more traditional strategies of broad-scale agricultural and industrial development, medicalization shifts the locus for development toward individual bodies. This signals the emergence of a development strategy that no longer focuses on influencing or coercing Third World states and elites, who are often overly responsive to national politics and institutional structures. This strategy instead focuses on daily surveillance and medicalization of individuals, far removed from centers of domination. It is a locally centered, dispersed approach to modernization—more a system of power in Foucault's sense than the type of global hegemony that is often discussed in international relations literature.[74] This strongly parallels the proliferation of knowledge around and disciplining of individual bodies that has accompanied modernization in the West; in many respects, this may even be the defining feature of modernity.[75] Foucault notes that such disciplining is achieved not merely through physical means but acts through rituals of surveillance, particularly self-surveillance. The increase of Western medical intervention in Third World countries should not, therefore, be seen as simply a way to improve health. It carries with it important cultural repercussions, one of which is modernization in the image of the West. This form of modernization has two immediate consequences. First, the focus on the individual can provide an entry for state-sanctioned violence.[76] It provides an internationally sanctioned means for controlling and coercing entire populations—one individual at a time. Second, although the focus on creating and supporting autonomy does provide many people with routes out of oppressive social, political, and cultural systems, it also provides the possibility for decontextualizing individuals when assessing their activities and needs in relation to welfare or violence—a trend within American culture that has given rise to a renewed rhetoric of fear and blame of the individual.[77]

Conclusion

This essay has provided a layered reading of overpopulation as it is articulated with other discourses and concerns in the United States. A historical reconstruction of policy discourse showed that such policies are enmeshed with knowledge about population/development links, rhetorics of environmental crisis, and the construction of various conflicting and reinforcing bodies.

This reading of overpopulation has sidestepped the usual debates over preferable methods of intervention or the need for greater or less involvement. The production of international population policy has instead been understood in a historical context as a site for international and domestic politics. Each of the shifts—from developmental policies, to "population control," to population neutrality, to "women's empowerment"—has been accompanied by shifts in linked discourses including forms of knowledge, political contexts, motivational rhetorics, institutions, and cultural constructions. Moving back and forth between these various arenas provides insight into the complexity of historical changes, and underscores the numerous continuities and sublimations that accompany any such shifts.

It can thus take a long journey through many arenas and interventions to see how decreasing Third World women's reproduction has become such a crucial site for working out numerous domestic, international, and global concerns—after all, the very people onto whom these concerns have been read are often not centrally in view. Before any interventions could be imagined, "the Third World woman," as an always silent, homogenized reproductive body, has been discursively constructed. This construction has been depoliticized and naturalized through the production of many scientific, political, and moral linkages between overpopulation and Third World women's bodies. The mapping provided in this essay explores such linkages and the production of the idea that "what is good for Third World women is good for the world" (or vice versa).

This discussion of overpopulation highlights the ways that, while often about poor, Third World women's bodies, these discourses are equally about Western selves, futures, and places in the world. Such a turn homeward should not be surprising. Scholars have interpreted most political action to be a struggle over identity, suggesting the need to understand how particular ensembles of individuals try to construct their own identities by engaging in international identity creation.[78] Such identity creation, in the case of overpopulation, is achieved by articulating discursive links between numerous different issues. The creation of Third World people as reproductive entities, or similarly, premodern pastoralists or unknowledgeable subsistence farmers, is achieved within the complex fields of discourse that circulate around overpopulation.[79] By looking at the struggles over exactly what this global role and place will be, we can begin to ask why overpopulation discourse takes on the form that it does. By looking at how these various discourses interact within political and institutional realms, we have begun to trace how overpopulation discourse is sustained across time and space.

Notes

1. I would like to thank Peter Taylor and Paul Edwards for extensive comments on several earlier drafts, and Susan DeLay for her inspiration, dialogue, and review of successive drafts of this essay.
2. Michel Foucault, *The Archaeology of Knowledge* (New York: Pantheon Books, 1972). See also Ernesto Laclau and Chantal Mouffe, *Hegemony and Socialist Strategy: Towards a Radical Democratic Politics* (New York and London: Verso, 1985), for a more thoroughly relational view of discourse.
3. Michel Foucault, *The History of Sexuality,* vol. 1, *An Introduction*, trans. Robert Hurley (New York: Vintage Books, 1980), 100.
4. "U.S. population policy," throughout this essay, refers to U.S. policy that is internationally focused. This sidelines both U.S. domestic population policy, which is tied strongly to immigration control and eugenics, and international population policy, which would necessitate a complex study in international relations. Although these various arenas are strongly related, their histories and actors are somewhat different.
5. Phyllis Tilson Piotrow, *World Population Crisis: The United States Response* (New York: Praeger Publishers, 1973); Jason L. Finkle and Barbara B. Crane, "The Politics of Bucharest: Population, Development, and the New International Economic Order," *Population and Development Review* 1, no. 1 (September 1975): 87–114; Jason L. Finkle and Barbara B. Crane, "Ideology and Politics at Mexico City: The United States at the 1984 International Conference on Population," *Population and Development Review* 11, no. 1 (March 1985): 1–28. Peter J. Donaldson, *Nature against Us: The United States and the World Population Crisis, 1965–1980* (Chapel Hill: University of North Carolina Press, 1990); Peter J. Donaldson, "On the Origins of the United States Government's International Population Policy," *Population Studies* 44 (1990): 385–99; Barbara Duden, "Population," in *The Development Dictionary: A Guide to Knowledge as Power*, ed. Wolfgang Sachs (Atlantic Highlands, N.J.: Zed Books, 1992); Marshall Green, "The Evolution of U.S. International Population Policy, 1965–92: A Chronological Account," *Population and Development Review* 19, no. 2 (June 1993): 303–21.
6. Green, "The Evolution of U.S. International Population Policy." Note that Green is far from an outside observer. He played a central role in the formation of population policy (and numerous antileftist interventions) throughout the 1970s. See Donaldson, *Nature against Us*, 72.

7. This shift parallels what Dennis Hodgson calls the shift to "orthodoxy" within demographic theory. As discussed in more detail later in this essay, "orthodoxy" describes the idea that rapid population growth slows or stalls development. Although "orthodoxy" thus provided a new notion of necessity, a belief in the feasibility of such programs seems to have its roots elsewhere.
8. Ansley J. Coale and Edgar M. Hoover, *Population Growth and Economic Development in Low-Income Countries* (Princeton, N.J.: Princeton University Press, 1958); Donaldson, "On the Origins of the United States Government's International Population Policy."
9. Donaldson, "On the Origins of the United States Government's International Population Policy," 387.
10. Piotrow, *World Population Crisis*; Donaldson, *Nature against Us*; Donaldson, "On the Origins of the United States Government's International Population Policy"; Green, "The Evolution of U.S. International Population Policy."
11. Finkle and Crane, "The Politics of Bucharest"; Finkle and Crane, "Ideology and Politics at Mexico City."
12. Barbara B. Crane and Jason L. Finkle, "The United States, China, and the United Nations Population Fund: Dynamics of U.S. Policymaking," *Population and Development Review* 15, no. 1 (March 1989): 23–59.
13. Jane Menken, "Introduction," in *World Population and U.S. Policy*, ed. Jane Menken (New York: Norton, 1986), 9–11. Julian Simon, who was the most influential of these advisers, provides a more detailed exposition of this position in *The Ultimate Resource* (New York: Pharos Books, 1981).
14. Finkle and Crane, "Ideology and Politics at Mexico City."
15. Ibid., 4.
16. Thomas W. Lippman, "Population Control Is Called a 'Top Priority' in Foreign Policy," *Washington Post*, January 12, 1994, A4.
17. Dennis Hodgson, "Orthodoxy and Revisionism in American Demography," *Population and Development Review* 14, no. 4 (December 1988): 541–69. The most recent stage is not discussed by Hodgson.
18. This theory was originally developed by theorists such as Frank Notestein and Warren Thompson in the 1930s and 1940s as a generalization from historical studies of European populations (ibid., 542).
19. For some of the early history of the discipline of demography, see Dennis Hodgson, "The Ideological Origins of the Population Association of America," *Population and Development Review* 17, no. 1 (March 1991): 1–34. For the solidification of its institutional base, see Hodgson, "Orthodoxy and Revisionism in American Demography." For a history of development discourse, see Arturo Escobar, *Encountering Development: The Making and Unmaking of the Third World*, Princeton Studies in Culture/Power/History, ed. Sherry B. Ortner, Nicholas B. Dirks, and Geoff Eley (Princeton, N.J.: Princeton University Press, 1995).
20. Escobar, *Encountering Development*, chapters 2–3. See also David Harvey, *The Condition of Postmodernity* (Cambridge, Mass.: Blackwell, 1989), for a discussion of the "overaccumulation crisis" in the Western industrialized countries, to which he attributes overseas expansion in the early postwar era.
21. Hodgson, "The Ideological Origins of the Population Association of America"; Frank W. Notestein, "Memories of the Early Years of the Association," *Population Index* 47, no. 3 (fall 1981): 484–88. On the growing significance of Asian countries, see Donaldson, "On the Origins of the United States Government's International Population Policy," 386.
22. Hodgson, "Orthodoxy and Revisionism in American Demography," 547.
23. Coale and Hoover, *Population Growth and Economic Development in Low-Income Countries*.

24. A summary of the orthodox position is provided by the National Research Council, *Population Growth and Economic Development: Policy Questions* (Washington, D.C.: National Academy Press, 1986). For some critiques, see Ozzie G. Simmons, *Perspectives on Development and Population Growth in the Third World* (New York: Plenum Press, 1988); Herman E. Daly, "Review: Population Growth and Economic Development: Policy Questions," *Population and Development Review* 12, no. 3 (September 1986): 582–85.
25. Hodgson, "Orthodoxy and Revisionism in American Demography," 557.
26. Simon, *The Ultimate Resource*.
27. Finkle and Crane, "Ideology and Politics at Mexico City."
28. Coale and Hoover, *Population Growth and Economic Development in Low-Income Countries*, 18–19.
29. National Research Council, *Population Growth and Economic Development*, 4.
30. Simmons, *Perspectives on Development and Population Growth in the Third World*.
31. Ibid. Escobar notes that such perspectives have still been "obliged to couch their critique in terms of the need for development. . . . it seemed impossible to conceptualize social reality in any other terms" (*Encountering Development*, 5). Discourse analytic approaches, such as those suggested by Escobar have not found their way into the policy arena.
32. See Green, "The Evolution of U.S. International Population Policy," for the history and Simmons, *Perspectives on Development and Population Growth in the Third World*, for more recent developments.
33. See, for instance, Susan Cotts Watkins, "If All We Knew about Women Were What We Read in *Demography*, What Would We Know?" *Demography* 30, no. 4 (1993): 551–78; Nancy E. Riley, "Challenging Demography: Contributions from Feminist Theory," Working Paper, Sociology Department, Bowdoin College; Susan Greenhalgh and Jiali Li, "Engendering Reproductive Policy and Practice in Peasant China: For a Feminist Demography of Reproduction," *Signs* (1995): 601–41.
34. Donaldson, *Nature against Us*; Piotrow, *World Population Crisis*. See Betsy Hartmann, *Reproductive Rights and Wrongs: The Global Politics of Population Control*, rev. ed. (Boston: South End Press, 1995), 111, for the official acceptance of coercion as possible U.S. policy.
35. For a clear expression of such linkages, see the 1965 congressional testimony of Thomas Ware, chairman of the Freedom from Hunger Foundation and CEO of International Minerals and Chemical Corporation. He linked the 1960s famines with population growth, technological ignorance, and Communist collectivization of farms, thus framing population control programs, the green revolution, and strong containment policies as a coherent package of programs (U.S. Congress, "World Population and Food Crisis," hearing, Senate Committee on Foreign Relations, 89th Congress, 1st session, June 29, 1965).
36. Peter J. Taylor and Raúl García-Barrios, "The Dynamics of Socio-Environmental Change and the Limits of Neo-Malthusian Environmentalism," in *Limits to Markets: Equity and the Global Environment*, ed. Mohamed Dore, Timothy Mount, and Henry Shue (Oxford: Blackwell, forthcoming), 15.
37. These include critiques of dehistoricization, globalization, depoliticization, ethics, ideology, and racism, among others. See, for instance, Mahmood Mamdani, *The Myth of Population Control: Family, Caste, and Class in an Indian Village* (New York: Monthly Review Press, 1972); H. S. D. Cole et al., eds., *Models of Doom: A Critique of the Limits to Growth* (New York: Universe Books, 1973); Paul Richards, "Ecological Change and the Politics of African Land Use," *African Studies Review* 26, no. 2 (June 1983): 1–72; Simmons, *Perspectives on Development and Population Growth in the Third*

World; Hartmann, *Reproductive Rights and Wrongs*; Susan Greenhalgh, "Controlling Births and Bodies in Village China," *American Ethnologist* 21, no. 1 (1994): 3–30; Taylor and García-Barrios, "The Dynamics of Socio-Environmental Change."

38. Raúl García-Barrios and Luis García-Barrios, "Environmental and Technological Degradation in Peasant Agriculture: A Consequence of Development in Mexico," *World Development* 18 (1990): 1569–85.
39. See, for example, Michael Watts, "Drought, Environment and Food Security," in *Drought and Hunger in Africa*, ed. Michael H. Glantz (Cambridge: Cambridge University Press, 1987), 171–211; Piers Blaikie, *The Political Economy of Soil Erosion in Developing Countries* (New York and London: Longman, 1985); Richards, "Ecological Change and the Politics of African Land Use."
40. Eric R. Wolf, *Europe and the People without History* (Berkeley: University of California Press, 1982); Richards, "Ecological Change and the Politics of African Land Use."
41. For a parallel discussion of the framing of certainty in global warming discourse, see Andrew Ross, "Is Global Culture Warming Up?" *Social Text* 28 (1991): 3–30; Peter J. Taylor, "How Do We Know We Have Global Environmental Problems?" (this volume).
42. Taylor, "How Do We Know We Have Global Environmental Problems?" 155.
43. Ross points out, however, that this does not necessarily rule out the usefulness of global environmental claims in forging liberatory global politics (Ross, "Is Global Culture Warming Up?").
44. Nancy Lee Peluso, "Coercing Conservation? The Politics of State Resource Control," *Global Environmental Change* 3, no. 2 (June 1993): 199–217.
45. Ann Stasia Anagnost, "Family Violence and Magical Violence: The Woman as Victim in China's One-Child Family Policy," *Women and Language* 11, no. 2 (1988): 16–22.
46. Soheir A. Morsy, "Safeguarding Women's Bodies: The White Man's Burden Medicalized," *Medical Anthropology Quarterly* 15, no. 1 (1991): 22.
47. The importance of bodies and behavior is highlighted by the centrality of fertility studies and surveys within demography and by the focus within policy arenas on individual control of births.
48. Hartmann, *Reproductive Rights and Wrongs*, 134.
49. John Cleland, "Marital Fertility Decline in Developing Countries: Theories and Evidence," in *Reproductive Change in Developing Countries: Insights from the World Fertility Survey*, ed. John Cleland and John Hobcraft (London: Oxford University Press, 1985), 225–27; Watkins, "If All We Knew about Women."
50. Donald P. Warwick, *Bitter Pills: Population Policies and Their Implementation in Eight Developing Countries* (Cambridge: Cambridge University Press, 1982), 34. See also Senator Alan Simpson's opening statements in U.S. Congress, "Foreign Aid Reform and S. 1096," Hearing, Senate Subcommittee on International Economic Policy, Trade, Oceans and Environment, February 22, 1994.
51. Note that although male bodies can also be implanted and sterilized, often with less trouble, most of these procedures have historically been carried out on women. See Hartmann, *Reproductive Rights and Wrongs*, 179–80; Christa Wichterich, "From the Struggle against 'Overpopulation' to the Industrialization of Human Reproduction," *Reproductive and Genetic Engineering* 1, no. 1 (1988): 21–30; Sonia Corría, *Population and Reproductive Rights: Feminist Perspectives from the South* (Atlantic Highlands, N.J.: Zed Books/DAWN, 1994), 25. This focus on women is often supported by a well-accepted postulate of population biology that a single fertile man can produce many babies, whereas a sterilized woman represents an absolute decrease in the fertility potential of a population. This perspective, of course, relies on an acultural model of human reproduction.
52. Anagnost, "Family Violence and Magical Violence."

53. Warwick, *Bitter Pills*, chapter 7; Mahmood Mamdani, *The Myth of Population Control: Family, Caste, and Class in an Indian Village* (New York: Monthly Review Press, 1972), chapter 7.
54. Faye Ginsburg documents a variant of this reading among domestic antiabortion activists who see abortion as a social tool for enforcing the worldview of the white, urban intelligentsia—a worldview that they associate with the breakdown of the family, relinquishing of "traditional" responsibilities, and sexual promiscuity. For these activists, abortion is tied up with cultural hegemony and the degradation of motherhood (Faye D. Ginsburg, *Contested Lives: The Abortion Debate in an American Community* [Berkeley: University of California Press, 1989]).
55. This can be seen clearly in the Reagan-era invocation of Chinese population control programs to highlight possible abuses of abortion.
56. This is evident in the domestic invocation of international abortion debates during the Cairo conference.
57. Susan Cohen, "The Road from Rio to Cairo: Toward a Common Agenda," *International Family Planning Perspectives* 19, no. 2 (June 1993): 61–66.
58. This is not a claim that such racial discourse is openly "racist," although racism certainly surfaces within this history.
59. Elizabeth Sobo discusses a secret National Security Council Memorandum from 1975 that put forth this position ("Why Washington Cares," *Progressive* [September 1990]: 28).
60. Al Gore, *Earth in the Balance: Ecology and the Human Spirit* (Boston: Houghton Mifflin, 1992), 308–9. One has to wonder whether he means "staggering" aesthetically, logically, or empirically. Congressman Anthony C. Beilenson, in 1993, did not even feel the need to mention such Western consumption patterns when he suggested that "the growing numbers of desperate poor will only accelerate the ferocious assault on the world's environment now underway in Africa, Asia, and Latin America" (U.S. Congress, "International Population Issues," Hearing, House Committee on Foreign Affairs, 103d Congress, 1st Session [September 22, 1993], 5). Conflicts over the relative importance of consumption rates versus population size have been recurrent in debates around overpopulation.
61. Quoted (without citation) in Frances Moore Lappé and Joseph Collins, *Food First: Beyond the Myth of Scarcity* (Boston: Houghton Mifflin, 1977), 6.
62. Garrett Hardin, "The Tragedy of the Commons," *Science* 162 (1968): 1246, 1247.
63. Ibid., 1247.
64. Paul R. Ehrlich, *The Population Bomb* (New York: Ballantine, 1968), 15.
65. Mamdani, *The Myth of Population Control*, 14.
66. Chandra Talpade Mohanty, "Under Western Eyes: Feminist Scholarship and Colonial Discourse," *Boundary 2* 12/13, no. 3/1 (1984): 333.
67. The potential for the return of such explicit race talk is suggested by a reenergized domestic focus on race. The relations between overpopulation and domestic race talk can be seen most clearly both in recent debates over immigration and in discussions of unwed teenage (black) welfare mothers.
68. It should be noted that most of this data is used in aggregate form. Also, there is debate about the overall usefulness of the fertility surveys at all. See Warwick, *Bitter Pills*, chapter 7.
69. Mohanty, "Under Western Eyes," 338.
70. This is not to say that eugenic notions do not still have currency in the general populace. The following letter to the editor was printed in response to an editorial against China's policy to use "extreme measures to avoid new births of inferior quality and heighten the standards of the whole population": "But we don't need to fear China for

a couple of hundred years. We need to fear ourselves because the so-called 'intelligent' are breaking the laws of nature; we are producing offspring who are breaking all laws. Eventually the laws of nature will prevail. One way or another, the strong will survive" (*St. Petersburg Times*, January 8, 1994, 19A).

71. Daniel J. Kevles, *In the Name of Eugenics: Genetics and the Uses of Human Heredity* (Berkeley: University of California Press, 1985); Hodgson, "The Ideological Origins of the Population Association of America."
72. Lars Bondestam and Steffan Bergstrom, *Poverty and Population Control* (New York: Academic Press, 1980); Farida Akhter, "The Eugenic and Racist Premise of Reproductive Rights and Population Control," *Issues in Reproductive and Genetic Engineering* 5, no. 1 (1992): 1–8; Sheldon Richman, "Letter: U.S. Betrays Heritage with Population Control," *Wall Street Journal*, June 15, 1993, A19.
73. That World Bank structural adjustment programs were privatizing health care in many of these countries at the same time (resulting in the destruction of many rural health clinics) seemed quite unrelated to this concurrent activity. See Hartmann, *Reproductive Rights and Wrongs*, 138–39, for the effects of structural adjustment programs on rural health care in the Third World.
74. Foucault, *The History of Sexuality*. For one take on hegemonic stability theory, see Stephen D. Krasner, *Structural Conflict: The Third World against Global Liberalism* (Berkeley: University of California Press, 1985).
75. Foucault, *The History of Sexuality*; Michel Foucault, *Discipline and Punish: The Birth of the Prison* (New York: Vintage Books, 1977).
76. Peluso, "Coercing Conservation?"
77. See, for instance, recent debates over welfare reform that target "welfare mothers" and "welfare cheats" as morally lax.
78. This idea draws on a very broad range of literature. See, for example, Edward W. Said, *Orientalism* (New York: Vintage Books, 1978); David Campbell, *Writing Security: United States Foreign Policy and the Politics of Identity* (Minneapolis: University of Minnesota Press, 1992); Alberto Melucci, *Nomads of the Present: Social Movements and Individual Needs in Contemporary Society*, ed. and trans. John Keane and Paul Mier (Philadelphia: Temple University Press, 1989). A critique of this approach is provided by Sankaran Krishna, "Review Essay: The Importance of Being Ironic: A Postcolonial View on Critical International Relations Theory," *Alternatives* 18 (1993): 385–417.
79. For an exploration of the latter, see Richards, "Ecological Change and the Politics of African Land Use."

How Do We Know We Have Global Environmental Problems? Undifferentiated Science-Politics and Its Potential Reconstruction

Peter J. Taylor

Introduction

More than a generation ago, scientists detected radioactive strontium from atomic tests in reindeer meat and linked DDT to the nonviability of bird eggs.[1] Ever since then, if not earlier, science has had a central role in shaping what count as environmental problems. During the 1980s, environmental scientists and environmentalists called attention in particular to analyses of carbon dioxide concentrations in polar ice, measurements of upper atmospheric ozone depletion, remote sensing assessments of tropical deforestation, and, most notably, projections of future temperature and precipitation changes drawn from computation-intensive atmospheric circulation models. This coalition of environmental activism and "planetary science" stimulated a rapid rise in awareness and discussion of global environmental problems.[2] A wave of natural and social-scientific studies has followed on the effects of global environmental change on vegetation and wildlife, agriculture, world trade and national economic viability, and international security. This science-centered environmentalism provides the first answer to the title question: we *know* we have global environmental problems because, in short, science documents the existing situation and ever tightens its predictions (or fills in its scenarios) of future changes. Accordingly, science supplies knowledge needed to stimulate and guide social-political action.

Science-centered environmentalism is, however, vulnerable to challenges and "deconstruction." Environmental problems, almost by definition, involve multiple, interacting causes. This allows one group of scientists to question the definitions and procedures of others, promote alternative explanations, cast doubt on the reliability of predictions, and emphasize the levels of uncertainty. In turn, people trying to make or influence policy often highlight the lack of scientific closure and the uncertainty.[3] After an initial honeymoon period, global climate modeling, estimates of biodiversity loss, and other studies of the implications of environmental change have become subject since the early 1990s to scientific and consequent political dispute.[4]

The purpose of this essay is not to add my own assessment of the reliability of global environmental science, the severity of the problems, or the appropriate framework for responding to the uncertainty of this science. Instead, building on the social studies of science, I propose a different interpretation of the special relationship between environmental science and politics, and then reflect on how such an interpretation could contribute to the potential reconstruction of environmental science-politics.[5]

The social studies of science have, over the past two decades, illuminated the social influences that shape what counts as scientific knowledge.[6] Truth or falsity of the science is rarely sufficient to account for its acceptance, either within science or, as will be an equally important concern here, within the political realm. Instead, to support their theories, scientists employ heterogeneous resources—equipment, experimental protocols, data, conventions of statistical analysis, citations, colleagues, the reputation of laboratories, metaphors, rhetorical devices, funding, publicity, and so on. Moreover, in this process of *heterogeneous construction*, establishing theory becomes just one aspect of scientific work.[7] Such a social studies of science perspective leads me to make three propositions, each confounding the answer given earlier to how we know we have global environmental problems.

1. In science, certain courses of action are facilitated over others, not just in the use or misuse of scientific results, but in how science is formulated in the first place—the problems chosen, categories used, relationships investigated, and confirming evidence required.[8] Politics—in the broad sense of courses of social action pursued or promoted—is not merely stimulated by scientific findings; politics is *woven into* the very fabric of science. In the case of environmental problems, we know they are *global* in part because scientists and political actors jointly construct them in global terms.[9]
2. In global environmental discourse, two allied views of politics—the moral and the technocratic—have been privileged. Both views of social action emphasize people's *common* interests in remedial environmental efforts, while at the same time steering attention away from the difficult politics that result from differentiated social groups and nations having different interests in causing and alleviating environmental problems.[10] We know *we* have global environmental problems, in part because the "we" referred to acts as if it were unitary and not a component of some highly differentiated population.[11]
3. Global environmentalism, whether as a framework for science or for political mobilization, is particularly vulnerable to deconstruction. Inattention to the national and localized political and economic dynamics of socioenvironmental change will ensure that scientists, both natural and

social, and the environmentalists who invoke their findings will be continually surprised by unpredicted outcomes, unintended conflicts, and unlikely coalitions. When environmental scientists (or some other group) attempt to focus on global environmental problems, to stand above the formation of such coalitions and conduct of such conflicts, and to discount their responsibility for the undesired outcomes of their policy proposals, they are more likely to reinforce the constraints on, rather than enhancing the possibilities of, engaged participants shaping interrelated, yet not common nor global, futures. In short, they know there are global environmental problems because they do *not* know most people do *not* have problems of a global nature.

To develop these propositions, I focus on one kind of environmental science, computer modeling of global environmental, resource, and climatic systems. I begin with a reconstruction and overview of the interwoven science and politics of the *Limits to Growth* (LTG) study of the 1970s.[12] This case, which should be familiar to most readers, is convenient because it illustrates the interweaving of science and politics clearly and allows me to introduce, in a somewhat exaggerated form, the moral and technocratic tendencies. From this beginning I make extensions to current studies of climate change and its human/social impacts, contrasting modeling work to analyses of environmental dynamics as *socio*environmental. This contrast is intended to speak also to other aspects of globalized, and more generally, undifferentiated, environmental discourse. Although I do not spell out the details of such extensions, it is in this spirit that I discuss examples indicating the vulnerability to deconstruction of such discourse. I conclude the essay by reflecting on my critique as a contribution to cultural politics.

Global Modeling, 1970s-Style

The Limits to Growth study was initiated by the Club of Rome, an elite group of Western businessmen, government leaders, and scientists, and conducted by system dynamics (SD) computer modelers at the Massachusetts Institute of Technology (MIT). The predictions from World 3, an SD model of the world's population, industry, and resources, were for population and economic collapse unless universal (coordinated, global-level) no-growth or steady-state policies were immediately established.

A major debate developed over the LTG study.[13] Environmentalists applauded the attention the LTG drew to the finiteness of the earth's resources, and many of them took up the steady-state economy as their major economic-environmental goal. Economists, however, strongly criticized the LTG's pessimism. Scarcity, signaled in price changes, they contended, would stimulate technological advance and thus push back the limits of available resources. From a different vantage point, many leftists and social justice-oriented progressives saw the LTG worldview as being insensitive to the needs of the poor and innocent of the realities of the penetration of multinational capital across the world.[14] Others, particularly those skilled in the methodology of systems analysis, pointed to weaknesses in the model's empirical basis, structure, and validation.[15]

Some of the technical objections were addressed in a subsequent Club of Rome-sponsored modeling effort, *Mankind at the Turning Point*.[16] This study disaggregated the world into ten regions and increased the detail of the model a thousandfold. Collapse was still predicted, but its timing and character would differ from region to region. By the time of this second report, however, the debate had cooled, a state of affairs that has been given divergent interpretations: the result of the unproductive polarization of pro-growth and antigrowth positions[17] or of incommensurable cultures/worldviews,[18] a decline in public environmental concern,[19] a shift toward greater specificity of discussion of environmental issues,[20] a quick rejection because the LTG's proposal for a steady-state economy threatened interests that were tied to economic growth and precipitated a "corporate veto."[21]

Despite the initial firestorm of criticism, the system dynamicists never conceded that their modeling was in error.[22] (Similarly, for many environmentalists the earth's finiteness became increasingly self-evident.) After the heated reaction to the LTG, the system dynamics group at MIT adopted a lower profile, but continued to use SD in a wide variety of modeling and educational projects,[23] most notably the explanation of broad modes of economic behavior—business cycles, inflation, and long waves (Kondratiev cycles). We can understand their continued belief in the validity of SD if we take another look at the construction of the LTG model of the world. While the system dynamicists were "doing science," they were also constructing interventions in their world. Both the representation of how the world works and the interventions proposed for improving it made each other seem more real and realizable. Moreover, we will see that the character of these representations/interventions was simultaneously moral and technocratic.

System dynamics, pioneered by Jay Forrester at MIT in the 1950s, was used first to model individual firms, then to explain urban decay, and, by the end of the 1960s, to uncover the dynamics of the whole world. The origin of SD in the modeling of firms has significance for the subsequent applications. Managers with whom Forrester had talked—recall that the LTG model and its predecessor models were developed at the Sloan School of Management at MIT—had observed repeated cycles of running up inventories, then laying off workers, and then once again accumulating a backlog of orders, adding labor, and increasing production, only to find themselves overcompensating and running up inventories again. Instead of attributing this cycle to the business cycle, Forrester concluded that the causes were endogenous to the firm. Each decision of management was rational, but, when the decisions were coupled together and incorporated the unavoidable time delays between setting a goal and fulfilling it, the overshoot-undershoot cycle resulted. Given that the undesirable behavior was caused by the interactions among different sectors of the firm, the firm's overall management could overcome the cycling only if there were a superintending manager in a position to override the decisions of managers in the separate sectors of the firm. For example, the sector managers could be instructed to keep larger inventories and respond more slowly to changes in the backlog of orders than they would otherwise prefer to do.

SD for firms set the pattern for the construction and validation of the subsequent urban, global, and other SD models. In general, the SD modeler does not rely primarily on series of recorded data, but instead invokes commonsense knowledge of how individuals work when they face a task with the usual information available. Computer games are often employed to convince players that they would not behave any differently from the people or other entities in the models.[24] Building on this commonsense validation of the separate decisions or rules in the model, SD then demonstrates that these locally rational decisions, when worked through time-delayed feedbacks in the system model, generate unanticipated, and undesired or pathological, outcomes.[25]

Using decision rules that look plausible to an individual, not only the LTG but almost all SD models exhibit undesirable cycles of positive feedback-based exponential growth and collapse. These cycles are difficult to overcome by adjusting the parameter values, even if set as high as economic or technological optimists would like. SD modelers infer that this behavior is intrinsic to the structure of the system modeled, to the arrangement of feedbacks, not their detailed specifications. The actions of some individuals *within* the system cannot override the structure, even if those individuals understand the system as a whole. Instead, a change in the structure is needed. In the case of the LTG "world system," however, unlike in firms, there is no superintending manager to enforce the required interrelated changes at this world level. Catastrophe is thus inevitable unless "everyone"—all people, all decision makers, all nations—can be convinced to act in concert to change the basic structure of population and production growth. In this fashion, SD models support either a moral response (everyone must change to avert catastrophe) or a technocratic response (only a superintending agency able to analyze the system as a whole can direct the changes needed). There is no paradox in my linking moral responses with technocratic ones; they are alike in attempting to bypass the political terrain in which different groups experience problems differently and act accordingly.

Does the nature of the politics indicated by some scientific results matter? Under a standard interpretation of science, it is no grounds for doubting the science. Forrester has argued that in order to address global questions, such as the "feasibility" of continued growth of the world's population, capital stock, and resource usage, global models are required.[26] One could, therefore, focus on the LTG's global models as science—do they provide an adequate account of the past and predictions of the future? However, following the interpretation that social actions are woven into the very formulation of science, I want to develop a stronger critique, one that addresses the LTG's science and politics simultaneously. If we consider how events would develop if population growth proved not, in Forrester's words, to be "feasible," a more politicized alternative to the LTG's analysis will become apparent.

Consider two hypothetical countries. Country A has a relatively equal land distribution; country B has a typical 1970s Central American land distribution: 2 percent of the people own 60 percent of the land; 70 percent own 2 percent. In other respects, these countries are similar: they have the same amount of arable land, the same population, the same level of capital availability and scientific capacity, and the same population growth rate, say, 3 percent. If we follow through the calculations of rates of population growth, food production increase, levels of poverty, and the like, we find that five generations before anyone is malnourished in country A, all of the poorest 70 percent in country B already are.[27] Food shortages linked to inequity in land distribution would be the likely level at which these poor people, and by implication most of the world's population, would first experience what others call "population pressure." In the LTG model, global aggregation of the world's population and resources obscured the fact that crises will not emerge according to a strictly global logic, much less in any global form as such. The *spatial* disaggregation in *Mankind at the Turning Point* does not resolve this issue. Land-starved peasants share nations, regions, and villages with their creditors, landlords, and employers. The sociopolitical responses of the peasants and, by extension, the ramifications of such local responses through national, regional, and international political and economic linkages will be (and already have been) *qualitatively* different from those highlighted by the LTG.[28]

This simple counterexample to global modeling does not tell us how to analyze the politics within localities, nations, regions, or the world, politics in which people contribute differentially to environmental problems. My point here is simply to highlight the politics of inequality excluded by the science of SD in its analysis of global limits to growth. The moral and technocratic emphasis is by no means a unique characteristic of the LTG study. My critique of the LTG's science-politics can be extended to the current globalization of environmental discourse.[29] Before doing so, let me first say a little more about the moral-technocratic alliance that such discourse generally presupposes.

In technocratic formulations, objective, scientific, and (typically) quantitative analyses are employed to identify the policies that society (or, in the case of the LTG, humanity) needs in order to restore order or ensure its sustainability or survival—policies to which individuals, citizens, and countries would then submit. In the LTG, these policies are deduced from the model structure, which is held to reveal a dynamic that the ordinary citizen, politician, or businessperson would not have recognized or specified. Moral formulations, in contrast, try to avoid coercion and rely on each individual making the change needed to maintain valued social or natural qualities of life. Yet in many senses the moral and the technocratic are allied. Both invoke the severity of the crisis and threat to our social order to command our attention. The solutions appeal to *common, undifferentiated* interests as a corrective to scientifically ignorant or corrupt, self-serving or naive governance. Moreover, appearances notwithstanding, special places in the proposed social transformations are reserved for their exponents—the technocrat as analyst/policy adviser; the moralist as guide, educator, or leader.[30]

Revealingly, the LTG report combined at numerous junctures managerial language and moral recruitment: "Until the underlying structures of our socio-economic systems are thoroughly analyzed, they cannot be *managed* effectively";[31] "The economic preferences of society *are [to be] shifted* more toward services";[32] "We cannot say with certainty how much longer *mankind* can postpone initiating deliberate control of *his* growth";[33] "The two missing ingredients are a realistic, long-term *goal* that can guide mankind . . . and the human *will* to achieve that goal."[34] In short, according to the LTG team, the global society needs management to achieve control; mankind as a whole, like an individual man, needs a goal and a will to change.

Global Modeling and Environmental Science Today

> We are moving into a period of chronic, global, and extremely complex syndromes of ecological and economic interdependence. These emerging syndromes threaten to constrain and even reverse progress in human development. They will be manageable—if at all—only with a commitment of

resources and consistency of purpose that transcends normal cycles and boundaries of scientific research and political action.[35]

Global climate models—or, more precisely, general circulation models (GCMs) of the atmosphere—have, especially since the hot, dry summer of 1988 in the United States, provided a new scientific basis for projections of imminent global environmental crisis. The actual modeling technique bears no similarity to system dynamics, but, as the diagnosis of environmental scientists William Clark and C. S. Holling illustrates, the language of the LTG lives on. More importantly for my argument, the science of global environmental change continues to reflect, and in turn reinforce, a tendency illustrated in extreme form by the LTG, namely, that toward moral-technocratic formulations of global environmental problems.[36] Two observations about contemporary research will serve to illustrate this point and to remind us of alternative formulations that, as in the LTG case, tend to be obscured by undifferentiated and globalized discourse.

First, consider the high premium that is placed on reducing uncertainty about physical processes in GCMs. To date, GCMs concur in predicting an average global warming, but the projected magnitude of the increase varies among the models. Second, at the level of regional predictions, larger uncertainties and inconsistencies among the GCMs are evident and need to be reduced. Indirect climatic feedbacks, creating new uncertainty, have been added to the research agenda.[37]

Tightening long-term projections or highlighting their severity is not, however, the only means by which policy responses to climate change could be catalyzed. As Michael Glantz has observed, extreme climate-related events, such as droughts, storms, and floods, already elicit sociopolitical responses that can be relatively easily studied.[38] Recent and historical cases of climatic-related "natural hazards" shed light on the impact of different emergency plans, investment in infrastructure and its maintenance, and reconstruction schemes. Policymakers, from the local level up, can learn "by analogy" from experience and prepare for future crises. Glantz's approach is valuable whether or not these crises increase in frequency (or are already increasing in frequency) as a result of global climate change. Instead of emphasizing the investigation of physical processes and waiting for uncertainty to be eliminated before action is taken from the top, this approach calls for systematic analysis of effective versus vulnerable institutional arrangements. Such discussion of specific, local responses to climate change has been occurring. Nevertheless, the vast majority of funds for global change research is currently being devoted to improving GCMs and allied climatic studies.

This dominance of physical climate research over institutional analysis points to a related issue, the hierarchy of physical science over the life sciences and social sciences. This hierarchy constitutes an environmental determinism: the physics and chemistry of climate change set the parameters for environmental and biological change; societies must then best adjust to the change in their environment. The hierarchy is evident in both the temporal sequence and the conceptual relationships of GCMs and other areas of environmental change research. GCM research began more than two decades ago. Building on the prominence given to GCMs in the late 1980s, a second tier of research arose that generates scenarios of agricultural, vegetation, and wildlife changes. This research models the interaction of projected temperature and precipitation changes with regional soils, watersheds, timing of snowmelts, wildfire susceptibility, coastal upwelling, and so on.[39] Following shortly after, a third tier of research was added that has been devoted to assessing the economic or security consequences of these biotic changes or of the more direct consequences of climate change, such as a rise in sea level.[40] Modes of geopolitical response to the global climate change threat then began to be discussed by political scientists.[41] Finally, social scientists and humanists began investigating popular understanding of global climate change, furnishing the bottom rung on the ladder from the hard and physical down to the soft and personal.[42]

Of course, global change researchers know that climate change is a social problem, since it is through industrial production, transport and electrical generation systems, and deforestation that societies generate greenhouse gases. Nonetheless, it is *physical change*—the mechanical and inexorable greenhouse effect—that is generally invoked to promote policy responses and social change, not the political and economic injustices of the present system.[43] Moreover, the research undertaken often belies the stated awareness of the social dimension of environmental problems. Natural scientists John Harte and his collaborators, for example, stated that "conservation policies designed without considering the role of existing institutions and societal responses to climate change are unlikely to be successful." Yet the same authors are open about their preference for models based on well-understood physical and biological *mechanisms* because these models are the ones that "work best for predicting change."[44] Not only do natural scientists favor their science over social analysis, but, in general, they have benefited from the prestige and funding that have flowed down from the high-status climate simulations. Despite Harte et al.'s caveat about the need to examine social responses, the politics implicit in their promotion of physical and biological models is technocratic. The prestige and funding given to their science bolsters scientists' confidence that political affairs can be influenced by technical knowledge without (or prior to) analysis of existing social arrangements.[45]

Again, the physical-natural-social-scientific hierarchy is not a *necessary* one for the construction of environmental problems and research. Over the last fifteen years, fields such as geography, anthropology, and international development studies have become increasingly sophisticated at analyzing environmental change as *socio*environmental change. Processes such as deforestation, drought, land degradation, and migration of "environmental refugees" are shown to be, in their causes and their effects, social and environmental at one and the same time.[46]

The social dynamics are most apparent on the economic level: resource distribution determines whether and *for whom* a bad year becomes a drought. Inequities in land tenure and rural political power ensure that the rural poor will exploit land vulnerable to erosion, migrate to carve new plots from the forest, or add to the margins of burgeoning cities well before the resources of their original locale are exhausted. Industrialization and other opportunities for off-farm income can result in insufficient labor remaining to keep up traditional conservation practices. Contrary to the conventional wisdom about the effect of population growth, environmental degradation can often be linked to labor *shortage*.[47]

Social dynamics are variable in interesting ways. In some areas, traditional practices have resisted disruption by linkage into global markets and instead contributed to environmental sustainability, whereas in other areas social organization has been rapidly restructured with significant environmental consequences.[48] To account for such differences one has to consider local particularity and historical contingency. Moreover, the local is not merely local, but "translocally" embedded, that is, influenced by institutions, processes, and activities well beyond the immediate locale. The local, in turn, can have distant ramifications—for example, the neglect of old terraces can lead to accelerated erosion and thus to siltation of reservoirs downstream.[49]

In a rich sense of the word "social," environmental problems invite social diagnosis and response. This will continue to be the case as climate change deepens and extends already existing crises. Many global environmental researchers consider themselves to have a worldview quite distinct from modelers,[50] yet, as long as this research remains dominated by physical and natural sciences and emphasizes social change as a response to environmental change, they underwrite, just as the LTG did, moral and technocratic responses. If they are unable to provide insight into the differentiated politics and economics of socioenvironmental change, what other responses *logically* are they leaving? In fact, because it omits any analysis of differentiated interests, undifferentiated discourse offers logically and conceptually no other standpoints for an environmentalist to take.[51]

Vulnerabilities of Undifferentiated Environmental Discourse

Global formulations of environmental issues have not gone unopposed. As I mentioned at the beginning of this essay, global climate modeling and studies of its implications have become subject to scientific dispute; policymakers, most notably in the United States, have used cracks in the scientific consensus and the unavoidable uncertainty about projections of future climate change to resist making new investments, regulations, and international treaties at this stage. In a complementary vein, influential economists have argued that the effects of climate change will be sufficiently gradual that adaptation mediated through the market and human migration is more cost-effective than rapid imposition of emission controls and other checks on economic growth.[52] From another angle, various Third World scholars and environmentalists have criticized Western analyses of emissions of gases contributing to the greenhouse effect. It is claimed that these exaggerate Third World contributions and fail to acknowledge the difference between the "survival emissions" of the Third World and the "luxury emissions" of the First World.[53]

The conceptual critique of the previous sections, however, leads me to identify different vulnerabilities of global environmental discourse, vulnerabilities that stem from different nations and differentiated social groups within nations having different interests in causing and alleviating environmental problems.[54] I should first make clear that my critique of global environmental discourse does not rely on the reader identifying wholeheartedly with global modeling, either in the form of the LTG study in the 1970s or with GCMs and current policy discussions motivated by them.[55] The LTG global models were not very detailed—even the SD modelers admitted this—and the possible policy responses could, therefore, be given only in outline. Obviously, the fashioning of the contributions of individual countries to a sustainable global system would involve considerable attention to their specific economic, political, and social conditions. Such specificity and messiness have been evident during the 1990s global environmental negotiations around the Rio and Cairo conferences and in other venues, such as the World Bank's Global Environmental Facility.[56]

My argument, then, is not that there are many governments that actually make policy as if moral-technocratic responses based on global modeling will be sufficient or successful.[57] Rather—and this makes the critique more general than modeling or global discourse—I want to draw attention to the surprises, from *unpredicted outcomes*, *unintended conflicts*, and *unlikely coalitions*, that tend to follow attempts, at whatever level, to discount the differentiated social dynamics and difficult politics of socioenvironmental change.

Let me illustrate these possibilities for surprise not with examples concerning global issues, but with four specific cases drawn from recent locally centered socioenvironmental studies.

(1) Jesse Ribot describes deforestation in Senegal, where there has long been a concern over depletion of forests exploited for wood fuel (including charcoal) and at the same time an awareness of problems of enforcing forest conservation policies.[58] The current policies might be characterized as a system of forest reserves to protect the shrinking forests, the establishment of extractive regions designated by official forestry agents, quotas and a limited season for charcoal production, and the organization and control of the production through permits, licenses, and, like most other rural sectors of the economy, a system of cooperatives. These policies follow a model, commonly advocated by international conservation organizations, of a neutral state acting in the best interests of the nation by ensuring the conservation of its natural resources.

The outcome has not been a decrease in the rate of deforestation. The quotas imposed are well below the urban demand, generating pressure for circumventing the official policies. This has engendered a myriad of means for producing outside the designated seasons and regions and for centralizing control of production in the hands of a few increasingly powerful individuals. To maintain this system, powerful individuals have a strong incentive to secure control over the institutions and operation of the different state bodies, which they do. The precolonial system of patrons and clients underwritten by political-religious authorities, which had been undermined under French colonialism, has been revived and strengthened under this forest policy regime. The newer, more disciplined organization of production and distribution has also intensified the deforestation, and the regulation by forestry agents of the official and unofficial systems has reduced the ability of local villagers to exclude production from their vicinity. In short, the policy of forest preservation, formulated without attention to the inequalities of social relations of Senegal, has produced an outcome clearly not intended by those who urge the protection of the world's remaining tropical forests.

(2) A long series of development schemes in the Gambia River Basin have failed to achieve their goals of intensifying rice production and reducing food imports. A scheme begun in the 1980s, based on sharecropping with specified planting, irrigation, and weeding requirements, has had more success, but, Judith Carney and Michael Watts show, it has also produced new struggles among men and women within peasant households.[59] Women have traditionally created and farmed rice in their own individual plots, separate from their work on household land. Men have, however, claimed house-

hold status for the plots established during rice projects, with the produce under their control. The current project, moreover, has increased women's labor, leading to women claiming individual status for their paddy, or, at least, remuneration for their work. In response, in some ethnic groups women have secured ownership, in others a share of the crop, in others nothing. Those receiving nothing have engaged in other gardening or trade activities or formed for-hire labor groups under the umbrella of their women's associations, which have become more active politically. The conflicts within the peasant household, which is clearly not a unitary entity, are not just based on force and formal property rights, but involve struggles over meaning, over representations of what men and women expect of each other. Moreover, the outcome of these struggles varies from one ethnic group to another. And the new labor coalitions were certainly not anticipated by the development planners. In this and other ways, domestic struggles extend and connect with other politics, that is, "dissent is manufactured."

(3) Richard A. Schroeder describes an analogous situation in the context of development projects promoting tree planting as a form of ecological stabilization in The Gambia.[60] After initial attempts to establish village woodlots failed, the emphasis shifted to fruit trees. These would provide quicker returns and distribution of the benefits could be more clearly defined. At the same time, however, the survival of these trees depended on their being planted within the borders of market gardens. It was assumed that the gardeners, who are women, would care for the young trees. Women, according to the prevailing development rhetoric, are more ecologically responsible. The market gardens, ironically, were the outcome of earlier development projects aimed to help women. The cash from these gardens compensated for declines of household income from the peanut crops managed by the men, but the women's superior earning power also led to conflict between husbands and wives. Further conflict has now ensued as men claim the fruit of the women's labor in the new orchards and the trees begin to shade out the gardens. Again, the conflicts within the peasant household are simultaneously struggles over meaning and struggles over land, labor, and production.

(4) In Nancy Lee Peluso's account of the coercive dimension of internationally endorsed conservation schemes, such as wildlife reserves in Kenya, she analyzes how environmentalists have been drawn into coalitions with the state and militarized institutions.[61] Many conservation schemes require or assume state control over natural resources, whereas this is often resisted by local peoples who have been gaining some of their livelihood from the resources in question—elephant tusks, game, products from the forests, and so on. Indeed, some of the very practices condemned by conservationists

arise as a consequence of previous State interventions. The Masai of Kenya, for example, began killing rhinoceroses and elephants (and later allegedly collaborating with ivory poachers) only after a long history of measures to restrict their traditional migratory cattle grazing and when it was clear that compensation in the form of jobs and income from tourism would never meet the original agreements. In response to poaching, the World Wildlife Fund and its partners not only endorsed, but provided aid to, programs equipping rangers with training, weapons, and other equipment. Peluso observes that, even before Richard Leakey's high-profile crackdown on poaching got under way in 1989, "the government was already using its mandate to protect and manage resources to assert [militarily] its authority" in a region near the Somali border. Conservation schemes have thus given the state and militarized institutions opportunities to gain more control of territory and peoples under a benevolent banner.

The kinds of surprises in these four examples are, I believe, the norm, not the exception, for environment, development, and conservation projects and policies. Some might conclude that no outside intervention should be attempted; there is always sufficient social complexity to produce unplanned consequences. This is not my point; in a world of interconnected economies, exchange rates, and structural adjustment, there is no such thing as nonintervention by outsiders. Moreover, if, in acknowledgment of differentiated social dynamics, we had highlighted the interest of corporations or dominant industrialized nations in environmentally destructive activities, we would still be far from capturing the difficult politics of socioenvironmental change. The lesson, instead, is that environmental scientists and activists need to take a position within the new coalitions and conflicts and work from there as the complex social processes unfold. To the extent that they discount their responsibility for the undesired outcomes of policy proposals, they are more likely to reinforce the constraints on, rather than enhance the possibilities of, engaged participants who are shaping interrelated, yet not common nor global, futures.

Reflection: The Construction of This Critique and the Potential Reconstruction of Undifferentiated Science-Politics

It should be clear that I oppose global environmentalism. I consider its science of undifferentiated dynamics to provide inadequate explanations, and policies based on such science not only unlikely to achieve their intended effects, but also likely to produce undesired ones. Instead of global environmentalism, I want to assert the need for a differentiated politics in all environmental discourse. Yet undifferentiated moral and technocratic discourse is pervasive, often being used comfortably by many who might not think of themselves as fitting the label technocrat or moralist.[62] How can an interpretation such as mine be expected to influence this state of affairs? When I reflected on this question, inconsistencies and other problems in the preceding sections struck me. Some additional work was needed, it seemed, to make sense of the construction of this critique and of its relation to the potential reconstruction of undifferentiated science-politics. Indeed, the self-conscious style of this concluding section follows from noting that interpreters of science have more or less assumed that their critique is cultural *politics*. This is not self-evident; the connection, I believe, needs exploration.

The most obvious role for this essay seems to be that critical science and/or culture interpreters, presumably making up a large fraction of this volume's readers, will appreciate the virtues of its critique of global environmentalism and proceed to disseminate and extend the critique. This reception would be pleasing; through some trickle-down or "diffuse out" process, a cadre of critics would make it harder for global environmental scientists and activists (hereafter "global environmentalists" or "GEers") to remain comfortable with undifferentiated discourse. But whether the critique of undifferentiated discourse comes from me or from a larger group of critics, if cultural politics is to extend beyond critique, the question remains: how does such an interpretation influence the original state of affairs?

GEers who have their attention drawn to the kind of critique contained in this essay might have an "aha!" experience, and from that point on reject globalized and undifferentiated discourse. Yet, to hope for that influence is inconsistent with the sociological perspective on science I have promoted here, namely, that interpretations and action, both scientific and social, are bound together, jointly reinforced by the formulation of problems, the tools available, the audiences being addressed and enlisted to act, the support (financial and otherwise) elicited, and so on (proposition 1).

In light of my promoting this sociological perspective, a different role for this essay might come from critical interpreters of science and/or culture appreciating and advancing this principle of interpretation. Again, questions of cultural politics would remain, but in this case the questions can be teased out further: What are critical interpreters supposed to do with and through interpretation that situates scientists as agents in a web of social resources? Can scientists be drawn into the audience for critical, situating interpretations? If so, how can their work build on (or in) such interpretations and lead to change in the original state of affairs? Let me explore these questions and in the process explicate as far as I can the *method* of interpretation that this essay is promoting.

Consider the combination of conceptual critique with practical critique. Conceptually, global environmental science is unable to provide insight into the differentiated politics and economics of socioenvironmental change; practically, policy or politics based on such science is vulnerable to surprises, from unpredicted outcomes, unintended conflicts, and unlikely coalitions. Evidently, by including a section on vulnerability of undifferentiated discourse, I thought that I needed to go further than conceptual critique of the science. In fact, I wanted my practical critique to say more than that the resulting policies are flawed and likely to result in undesired effects. By emphasizing surprises of the kind "environmentally motivated projects lead to household conflict and breakdown" or "conservation leads to coercive environmentalist-state coalitions," I was raising the level of polemic, hoping to prod GEers morally—surely they would want to change their ways if they considered these kinds of consequences.[63]

Looking back, it is clear that even before I mentioned the striking surprises, propositions 1 and 2 foreshadowed this moral prodding. By arguing that certain politics (here, the moralistic and technocratic tendencies) and also the science that facilitates them are not dictated by the nature of reality, it follows that scientists and other social agents can choose whether or not to contribute to such science-politics. They are thus partly and jointly responsible for the consequences. Proposition 3 then built on that: in order to urge GEers to acknowledge that responsibility I wanted to stress that their science-politics does have significant consequences; policies based on undifferentiated analyses make unintended effects and undesirable surprises inevitable.

Yet why should one expect conceptual, practical, and moral critique, even when combined, to provide an adequate way of shifting discourse? Countervailing pragmatic and practical reasons can be readily identified that help us understand why in this case GEers might be attracted to moral-technocratic politics: *(a)* moralistic recruitment to a cause and appeals to universal interests can be effective as political tactics—human rights campaigns in times of severe political repression demostrate that; *(b)* more generally, political mobilization usually depends on stressing commonality of interests and playing down differences—since the Apollo space photographs of the earth and, more recently, the end of the Cold War, it has become popular to speak of the common future of earth's inhabitants; *(c)* for scientists, a technocratic outlook is an understandable orientation—they would rather apply their special skills and institutional location as best as they can to benefit society, than expend energy in political organizing for which they are less likely to have experience or aptitude; and *(d)* it is difficult to communicate well with others engaged in a discourse without using the common language, which, as I have noted in the case of global environmental discourse, makes extensive use of the terms of management and/or moralistic recruitment and education. In short, in an extension of propositions 1 and 2, many people know that we have global environmental problems because *their* institutional, linguistic, and social location facilitates global discourse and the tendencies to moral-technocratic politics.

Having pointed to the *practical* facilitations of the moralistic and technocratic tendencies, I cannot expect these tendencies to be undermined by a mere counterinterpretation, that is, something working mostly on an *intellectual and textual* level. One needs to understand and counteract the wider sources of the popularity of undifferentiated discourse in order to oppose it.[64] Indeed, recall the social studies of science view of the heterogeneous construction of scientific activity: "to support their scientific theories and other work scientists employ heterogeneous resources." A straightforward extension of this perspective would be to say that GEers employ heterogeneous resources to support their global environmentalist activity. This perspective, especially when combined with the previous sections' emphasis on differentiated analysis, could have led me to analyze the *multifaceted* ways that politics become woven into environmental knowledge.

The logical extension of my framework would have been to investigate particular cases of environmental knowledge making, and in light of the diverse facilitations observed, to contribute to building conditions favorable to alternative science and politics.[65] Indeed, this is an ideal I will return to at the end of this essay. In the absence of the detailed work on particular cases, proposition 2 could, by analogy, lead critical readers to interpret generalizations such as propositions 2 and 3 themselves as my attempt to make space for social studies of science in environmental discourse while avoiding dealing with the particularities, messiness, and other difficulties of achieving change (here, the change to be achieved would be in environmental analysis and policy).[66]

Let me acknowledge these inconsistencies. I could, by way of excusing myself, point to the character of this volume, the need to avoid specialized discussion if I were to reach readers from many disciplines, the limitations on essay length, the constraints of devoting time to my primary research and to other commitments, and so on. For all these reasons, it would not have been possible or appropriate to present any differentiated, locally centered, translocal analyses of the politics of environmental knowledge making.[67] I think, however, that I can proceed in a way that is more helpful and generally applicable than this somewhat defensive response.

There is a positive method of interpretation of science-politics implied in my essay that centers around *heuristics*. As I think of them, heuristics are propositions that stimulate, orient, and guide our inquiries; they are useful provided that we remember that they break down when too much weight is given to them. Let me tease out the different ways that propositions 1–3 can be applied heuristically to contribute to reflection on and intervention within the politics of knowledge.

One point of entry is to begin from what I will call heuristic 1: assume that scientists seek logical consistency among their different ideas. This allows me to use heuristic 2: identify how the form of different scientific theories logically admits particular forms of intervention in the world. In this spirit, I stated, for example: "because it omits any analysis of differentiated interests, undifferentiated discourse offers logically and conceptually no other standpoints [other than moral/technocratic politics] for an environmentalist to take." Heuristic 3 then follows: use the logic of the science-policy connection to tell a generalized story with a moral, hoping that the moral is an effective prod for some GEers to seek changes in their science and politics. In this spirit, I described the scenario of the two islands and the possibility of research on differentiated socioenvironmental dynamics, and then amplified this by highlighting the undesirable surprises that follow from undifferentiated analyses.[68]

Note that heuristic 3 leaves to each GEer the task of mobilizing the heterogeneous resources needed to effect change in their own particular circumstances. In other words, although the moral appeal can have some rhetorical value, it does not identify in any detail the materials to use in bringing about change. Note also that, without heuristic 1, the logical connection between representations and interventions (using heuristic 2) would not be very telling (i.e., heuristic 3 would not have much rhetorical power). Scientists would feel free to persist in developing their undifferentiated models despite someone's pointing out the logical connection with policy interventions and outcomes of which they disapproved.

Suppose, instead, we lessen the weight placed on heuristic 1 and simply assume that scientists are somewhat constrained by issues of logical consistency in the formulation of their science and their policy proposals. This revised version leads to other heuristics that begin to expose more of the diverse facilitations involved in environmental knowledge making and policy making:

Heuristic 4: frame the logical extension of scientific theories into practice and policy as an "accusation"—for example, "This science supports a moral and technocratic politics." The intention would be to provoke responses that might reveal more of the diverse practical as well as intellectual resources that the particular GEers are employing. Responses elicited could include: "I am not a technocrat"; "I do not condone coercive conservation—I know that in my heart"; "By what framework can you interpret my motives [differently from the way I do]?" Some of these responses might stop the exchanges dead in their tracks, but if the conversation continues it becomes possible to ask questions such as: "What do you think about the observation that global environmental scientists have, in practice, shifted readily between the language of enlisting the readers to change and the language of management and control?" "How would you incorporate unequal agents and the dynamics of differentiation into your analyses?"

Whether by means of such dialogue or working with written materials, heuristic 5 can be employed: examine the ways that particular GEers address the logical extensions; that is, using these extensions as starting places or "null hypotheses," consider how the particular GEers address differentiated politics (e.g., ask who makes the changes they propose), how they position themselves (e.g., ask who are their sponsors, allies, audiences), how they use science (e.g., ask what are their preferred categories, data sources, mathematical and computer tools, etc.).

These heuristics can lead us a little distance toward analyzing the heterogeneous resources drawn into the construction of any scientific-political activity. From the point of view of cultural politics, the desirability of such an analysis is that it would enable the interpreter or reader to identify multiple sites of potential intervention—none decisive on its own (they need to be *linked* to lead to effective change), but each more doable than moral-technocratic prescriptions for change (heuristic 6). To move further toward this ideal, however, requires heuristics and "shifts of positioning" that go far beyond this essay.[69]

Reflection on my critique has led me to reformulate the three propositions of this essay in terms of a set of heuristics with which to begin to expose more of different and unequal GEers' actual, particular constructions of globalized environmental discourse. Yet the original propositions can still be read, whether one agrees with them or not, as attempts, like the frameworks they critiqued, to cut through the unequal and heterogeneous practical and conceptual facilitations of science and political mobilization. I have not eliminated this tension; by either reading, the interpretation of science-politics introduced in this essay provides some important resources to probe and intervene in the networks that GEers build to support their science-politics. But cultural politics of science should call for practical engagement in these processes, not just critical interpretation. Between undifferentiated science-politics and its interpretation, and between interpretation and reconstruction lies a world of difference—and of ongoing differentiation.

Notes

1. The first three sections of this essay are drawn, with revisions, from Peter J. Taylor and Frederick H. Buttel, "How Do We Know We Have Global Environmental Problems? Science and the Globalization of Environmental Discourse," *Geoforum* 23, no. 3 (1992): 405–16. I gratefully acknowledge Fred Buttel's collaboration in the original paper and the permission of *Geoforum* for me to use excerpts in this essay. The comments of Paul Edwards, Saul Halfon, and David Takacs, together with the bibliographic advice of Paul Edwards and Clark Miller, have helped me revise the original paper. Jesse Ribot's feedback on my rendering of his research (note 58) and Kim Berry's summary of the Carney and Watts study (note 59) have also been valuable.
2. In its earlier years, modern environmentalism promoted other science-based, global formulations of environmental problems—for example, Ehrlich's "population bomb" (which built on population biology) and Meadows et al.'s (1972) "limits to growth" (derived from the application of system dynamics to population and resources). See Paul R. Ehrlich, *The Population Bomb* (New York: Ballantine, 1968); and Donella Meadows, Dennis Meadows, Jørgen Randers, and William Behrens, *The Limits to Growth* (New York: Universe Books, 1972). The current manifestation is, however, more broad-based.
3. Sheila S. Jasanoff, "Science, Politics, and the Renegotiation of Expertise at EPA," *Osiris* 7 (1992): 1–23; Simon Shackley and Brian Wynne, "Representing Uncertainty in Global Climate Change Science and Policy: Boundary-Ordering Devices and Authority," *Science, Technology, and Human Values* 21, no. 3 (1996): 275–302.
4. Compare the treatments in scientific journals, for example, by Reid Bryson, "Will There Be a Global 'Greenhouse' Warming?" *Environmental Conservation* 17 (1990): 97–99; William E. Reifsnyder, "A Tale of Ten Fallacies: The Skeptical Enquirers' View of the Carbon Dioxide/Climate Controversy," *Agricultural and Forest Meteorology* 47 (1989): 349–71; Charles C. Mann, "Extinction: Are Ecologists Crying Wolf?" *Science* 253 (1991): 736–38; Paul R. Ehrlich and Edward O. Wilson, "Biodiversity Studies: Science and Policy," *Science* 253 (1991): 758–62; William W. Kellogg, "Response to Skeptics of Global Warming," *Bulletin of the American Meteorological Society* 72 (1991): 499–511; and the Marshall Institute, *Scientific Perspectives on the Greenhouse Problem* (Washington, D.C.: Marshall Institute, 1989). In the popular press, compare Anil Agarwal and Sunita Narain, "Global Warming in an Unequal World: A Case of Environmental Colonialism," *Earth Island Journal* (spring 1991): 39–40; Anonymous, "Species Galore: Avoiding Extinctions Should Not Be an Overriding Goal for Environmentalists," *Economist* 320, no. 7724 (1991): 17; and William D. Nordhaus, "Greenhouse Economics: Count Before You Leap," *Economist* 316, no. 7662 (1990): 21–24.

   In the case of global climate science, partly in response to the challenges and partly in response to the "untidy political processes" involved in forming policy, many scientists have withdrawn into more neutral positions than those taken from the mid-1980s to the early 1990s; see Sonja A. Boehmer-Christiansen, "A Scientific Agenda for Climate Policy?" *Nature* 372 (1994): 400–402, and Paul N. Edwards, "Global Comprehensive Models in Politics and Policymaking," *Climatic Change* 32, no. 2 (1996): 149–61.

5. In a similar spirit, sociologists of science Simon Shackley and Brian Wynne argue that the "criteria for 'good science' with respect to GCMs [atmospheric circulation models] are not being defined solely from within science itself, but are in part products of the interactions of science with other domains, particularly the policy world" ("Global Climate Change: The Mutual Construction of an Emergent Science-Policy Domain," *Science and Public Policy* 22, no. 4 [1995]: 218–30). See also Sheila S. Jasanoff and Brian Wynne, "Scientific Knowledge and Decision Making," in *State of the Art Report on Climate Change*, ed. Steve Rayner (Richland, Wash.: Batelle—Pacific Northwest Laboratories, 1997).
6. See, for example, Randall Collins and Sal Restivo, "Development, Diversity, and Conflict in the Sociology of Science," *Sociological Quarterly* 24 (1983): 185–200; S. Leigh Star, "Introduction: The Sociology of Science and Technology," *Social Problems* 35, no. 3 (1988): 197–205; Steve Woolgar, *Science: The Very Idea* (London: Tavistock, 1988).
7. John Law, "On the Methods of Long-Distance Control: Vessels, Navigation and the Portuguese Route to India," in *Power, Action, Belief*, ed. John Law (London: Routledge and Kegan Paul, 1986), 234–63; Bruno Latour, *Science in Action: How to Follow Scientists and Engineers through Society* (Milton Keynes: Open University Press, 1987); Peter J. Taylor, "Building on Construction: An Exploration of Heterogeneous Constructionism, Using an Analogy from Psychology and a Sketch from Socio-Economic Modeling," *Perspectives on Science* 3, no. 1 (1995): 66–98.
8. Taylor, "Building on Construction"; Peter J. Taylor, "Technocratic Optimism, H. T. Odum, and the Partial Transformation of Ecological Metaphor after World War II," *Journal of the History of Biology* 21, no. 2 (1988): 213–44; Peter J. Taylor, "Re/constructing Socio-Ecologies: System Dynamics Modeling of Nomadic Pastoralists in Sub-Saharan Africa," in *The Right Tools for the Job: At Work in Twentieth-Century Life Sciences*, ed. Adele Clarke and Joan Fujimura (Princeton, N.J.: Princeton University Press, 1992), 115–48.
9. Although the term "global" is used by many as synonymous with "international" or "transnational," I use "global" in this essay strictly in the sense of "the world as a whole." My analysis is intended to apply to international environmental research only if it (1) slips into, or rides the coattails of, globalized (in the strict sense) discourse, or (2) treats agents as undifferentiated (see note 11). Notes 29, 36, 45, 50, 51 also help to maintain these distinctions and focus.
10. See also Diane M. Liverman, "Vulnerability to Global Environmental Change," in *Understanding Global Environmental Change: The Contributions of Risk Analysis and Management*, ed. Roger E. Kasperson et al. (Worcester, Mass.: Earth Transformed Program, Clark University, 1990), 27–44.
11. Later in the essay I subsume global environmental discourse within a larger class, namely, environmental discourse for which the agents are undifferentiated. The form of my critique is intended to apply to this larger class.
12. Meadows et al., *The Limits to Growth*. See also Taylor, "Re/constructing Socio-Ecologies," for a more detailed sociological analysis of a closely related system dynamics environmental modeling project.
13. See the useful summary in Robert McCutcheon, *Limits of a Modern World* (London: Butterworths, 1979).
14. Francis Sandbach, "The Rise and Fall of the Limits to Growth Debate," *Social Studies of Science* 8 (1978): 495–520; David L. Sills, "The Environmental Movement and Its Critics," *Human Ecology* 3 (1975): 1–41.

15. See especially H. S. D. Cole et al., eds., *Models of Doom: A Critique of the Limits to Growth* (New York: Universe Books, 1973).
16. Mihajlo D. Mesarovic and Eduard Pestel, *Mankind at the Turning Point* (New York: Dutton, 1974).
17. Lincoln Gordon, "Limits to the Growth Debate," *Resources* 52 (summer 1976): 1–6.
18. Brian P. Bloomfield, *Modelling the World: The Social Constructions of Systems Analysts* (Oxford: Blackwell, 1986).
19. Sandbach, "The Rise and Fall of the Limits to Growth Debate."
20. Craig Humphrey and Frederick H. Buttel, *Environment, Energy, and Society* (Belmont, Calif.: Wadsworth, 1982), 110.
21. Frederick Buttel, Ann Hawkins, and Alison Power, "From Limits to Growth to Global Change: Constrasts and Contradictions in the Evolution of Environmental Science and Ideology," *Global Environmental Change* 1, no. 1 (1990): 57–66.
22. Donella Meadows, Dennis Meadows, Jørgen Randers, and William Behrens, "A Response to Sussex," in Cole et al., *Models of Doom*, 216–40; Bloomfield, *Modelling the World*.
23. For example, Jay W. Forrester, "Educational Implications of Responses to System Dynamics Models," in *World Modeling: A Dialogue*, ed. C. West Churchman and Richard O. Mason (New York: American Elsevier, 1976), 27–35.
24. John Sterman, "Testing Behavioral Simulation Models by Direct Experiment," *Management Science* 33 (1987): 1572–92.
25. The science here is not exceptional; all model making ultimately depends on certain assumptions being accepted on the basis of their plausibility, rather than on tight correspondence with empirical data. See Peter J. Taylor, "Revising Models and Generating Theory," *Oikos* 54 (1989): 121–26. Economists, in particular, are unapologetic about this; see Milton Friedmann, *Essays in Positive Economics* (Chicago: University of Chicago Press, 1953).
26. Forrester, "Educational Implications"; see also Meadows et al., "A Response to Sussex," 238.
27. For a detailed analysis of this issue in a nonhypothetical case, El Salvador, see William D. Durham, *Scarcity and Survival in Central America: Ecological Origins of the Soccer War* (Stanford, Calif.: Stanford University Press, 1979). The hypothetical scenario is derived originally from John Vandermeer, "Ecological Determinism," in *Biology as a Social Weapon*, ed. Science for the People (Minneapolis: Burgess, 1977), 108–22. See also Peter J. Taylor and Raúl García-Barrios, "The Dynamics of Socio-Environmental Change and the Limits of Neo-Malthusian Environmentalism," in *Limits to Markets: Equity and the Global Environment*, ed. Mohamed Dore, Timothy Mount, and Henry Shue (Oxford: Blackwell, 1996).
28. William W. Murdoch, *The Poverty of Nations* (Baltimore: Johns Hopkins University Press, 1980); Amartya K. Sen, *Poverty and Famines* (Oxford: Oxford University Press, 1981); Carol A. Smith, "Local History in Global Context: Social and Economic Transitions in Western Guatemala," *Comparative Studies in Society and History* 26, no. 2 (1984): 193–228. Similarly, pollution, which was modeled in the LTG as an aggregate world level, is differentially distributed by class and race; see Robert Bullard, *Dumping in Dixie: Race, Class and Environmental Quality* (Boulder, Colo.: Westview Press, 1994).
29. The politics of attempting to bypass difficult politics, which is characteristic of modeling and systems approaches, is also evident in many other areas of environmental discourse. Consider: neo-Malthusians' use of aggregate population statistics (Paul R. Ehrlich and Anne H. Ehrlich, *The Population Explosion* [New York: Simon and Schuster, 1990]); deep ecology's emphasis on the need for a biocentric ethic (George Brad-

ford, "How Deep Is Deep Ecology? A Challenge to Radical Environmentalism," *Fifth Estate* 22, no. 3 [1987]: 3–64); conservation biology's celebration of endangered species and losses in biodiversity (Ehrlich and Wilson, "Biodiversity Studies," 758–62); and the moderate tone of sustainable development's language (Sharad Lélé, "Sustainable Development: A Critical Review," *World Development* 19, no. 6 [1991]: 607–21) in contrast with earlier analyses of dependency and necessary underdevelopment (Paul Cammack, "Dependency and the Politics of Development," in *Perspectives on Development: Cross-Disciplinary Themes in Development Studies*, ed. P. F. Leeson and M. Martin Minogue [Manchester: Manchester University Press, 1988], 89–125).

30. Taylor, "Technocratic Optimism."
31. Meadows et al., *Limits to Growth*, 181; emphasis added.
32. Ibid., 163; emphasis added.
33. Ibid., 183; emphasis added.
34. Ibid., 184; emphasis added.
35. William C. Clark and C. S. Holling, "Sustainable Development of the Biosphere: Human Activities and Global Change," in *Global Change*, ed. Thomas F. Malone and Juan G. Roederer (Cambridge: Cambridge University Press, 1985), 477. See also William C. Clark, "Managing Planet Earth," *Scientific American* 261, no. 3 (1989): 47–54.
36. It seems very difficult for anyone to engage in globalized environmental discourse and enlist others in their point of view without slipping into the languages of management and/or moralistic recruitment and education. This was brought home to me in reviewing writings more recent than those of Meadows et al. and Clark and Holling. Quotes from two additional sources illustrate how language that is familiar and well-meaning partakes of these two tendencies (emphasis added in all quotes).

In the discussion papers and notes circulated in preparation for a volume on equity and sustainability—Philip B. Smith et al., eds., *The World at the Crossroads: Towards a Sustainable, Equitable and Liveable World* (London: Earthscan, 1994)—I read of a call for "a total picture of the world" and "*rechannel*[ing] activity into sustainable forms," phrases that conjure up the hubris of a technocrat. Moralistic language was, however, more pervasive. Recruitment to the cause of responding to "our" common prospect was implied in the recurrent use of "we," "our culture," "our existence," "humanity," and in phrases such as "*our* built-in limitations of perception," "time available for *us* to change *our* ways." One paper discussed whether "society could *be changed* quickly enough," basing its claims around behavioral characteristics supposedly given to humans by their evolutionary history; that is, we are all fundamentally alike, being members of the same species. Individual behavior and social dynamics were often expressed in the same undifferentiated terms, with individual metaphors used for social ideas and without mention of any structure between the individual and society: "Will humankind take the fork leading to disaster or . . . to survival?" Does society have the "*will* to alleviate poverty"? "Affluent societies can choose," despite the "perennial foot-dragging of the establishment." "Individuals vary, [therefore] societies vary." (In the final published volume, quotations such as these were accompanied by much more attention to stratification within and among nations [but not the dynamic interactions among strata; see note 45], and the technocratic currents were less apparent.)

The editorial for the journal *Conservation Biology*—Gary K. Meffe, Anne H. Ehrlich, and David Ehrenfeld, "Human Population Control: The Missing Agenda," *Conservation Biology* 7, no. 1 (1993): 1–3—speaks of conservation biologists "possess[ing] the professional responsibility to *teach* humankind about the perils" (2) of continued population growth, "having the obligation to provide *leadership* in addressing the human population problem and developing solutions" (ibid.), and being able to "help promote policies to *curb* rapid population growth" (3). "The population problem is stunningly clear and ought to be beyond denial" (2). "The human species ignores or denies" the impending calamity (ibid.). (Presumably those who draw attention to the population problem are excused from this species collectivity.) A brief mention of the "critical importance . . . of educating and empowering women" (3) in the next-to-last paragraph hints that all people might not be equally responsible, but the concluding paragraph returns to the dominant undifferentiated formulation: "Action is needed from *everyone*, at every turn . . . [in the cause of] human population control. Life itself is at stake" (ibid.).

Language does not, however, stand on its own and readers should not forget the conceptual argument about the undifferentiated dynamics entailing moralistic or technocratic responses.

37. Daniel Lashof, "The Dynamic Greenhouse: Feedback Processes That May Influence Future Concentrations of Atmospheric Trace Gases," *Climatic Change* 14 (1990): 213–42.
38. Michael Glantz, ed., *Societal Responses to Regional Climatic Change: Forecasting by Analogy* (Boulder, Colo.: Westview Press, 1989).
39. Peter Gleick, "The Implications of Global Climatic Change for International Security," *Climatic Change* 15, nos.1–2 (1989): 309–25; John Harte, Margaret Torn, and Deborah Jensen, "The Nature and Consequences of Indirect Linkages between Climate Change and Biological Diversity," in *Consequences of the Greenhouse Effect for Biological Diversity*, ed. Robert L. Peters and Thomas E. Lovejoy (New Haven: Yale University Press, 1992): 325–43; Yvonne Baskin, "Ecologists Put Some Life into Models of a Changing World," *Science* 259 (1993): 1694–96.
40. United States Department of Energy, "The Economics of Long-Term Global Climate Change: A Preliminary Assessment" (1990); Thomas F. Homer-Dixon, "On the Threshold: Environmental Changes as Causes of Acute Conflict," *International Security* 16, no. 2 (1991): 76–116.
41. See special issues of *Policy Studies Journal* 19 (spring 1991) and *Evaluation Review* 15 (February 1991).
42. Willett Kempton, "Lay Perspectives on Global Environmental Change," *Global Environmental Change* 1 (1991): 183–208. For a more socially interpretative account of planetary science, see Andrew Ross, "Is Global Culture Warming Up?" *Social Text* 28 (1991): 3–30.
43. See, for example, B. L. Turner et al., "Two Types of Global Environmental Change: Definitional and Spatial-Scale Issues in Their Human Dimensions," *Global Environmental Change* 1, no. 1 (1990): 15.
44. Harte et al., "The Nature and Consequences of Indirect Linkages," 339, 341.
45. The complementary moral politics of global climate and environmental change researchers requires a small qualification. Almost all commentators acknowledge that there are rich and poor groups, peoples, and nations; that the rich consume more per capita; and that poverty may compel the poor to "mine" their resources. Acknowledging the statistics of inequality does not, however, constitute an analysis of the *dynamics* of inequality. In the absence of serious intellectual work—conceptual and empirical—heartfelt caveats about the rich and the poor do not substantially alter the politics woven into this research. This discourse has simply separated the moral appeals into two uniform audiences: all the poor must change (e.g., practice family planning); all the rich must change (e.g., reduce consumption). It should be noted that some social-science analysts who describe phenomena such as the poor mining their resources recognize that the resource degradation makes sense only in the context of differentiated dynamics, that is, of the exploitation of the poor. Nevertheless, they steer away from stating this out of deference to "policy palatability" or as an accommodation to the

dominance of neoliberal economics since the early 1980s; see Frederick H. Buttel and Peter J. Taylor, "Environmental Sociology and Global Environmental Change: A Critical Assessment," *Society and Natural Resources* 5 (1992): 211–30.

46. Equivalently, ecology is *political ecology*. See Michael Watts and Richard Peet, eds., *Environment and Development. Special Double Issue of Economic Geography* 69, nos. 3–4 (1993): 227–448; Peter J. Taylor and Raúl García-Barrios, "The Social Analysis of Ecological Change: From Systems to Intersecting Processes," *Social Science Information* 34, no. 1 (1995): 5–30; Rod Neumann and Richard Schroeder, "Manifest Ecological Destinies: Local Rights and Global Environmental Agendas," *Antipode* 27, no. 4 (1995): 321–448.
47. Raúl García-Barrios and Luis García-Barrios, "Environmental and Technological Degradation in Peasant Agriculture: A Consequence of Development in Mexico," *World Development* 18, no. 11 (1990): 1569–85; Taylor and García-Barrios, "The Dynamics of Socio-Environmental Change."
48. Peter Little, "Land Use Conflicts in the Agricultural/Pastoral Borderlands: The Case of Kenya," in *Lands at Risk in the Third World: Local Level Perspectives*, ed. Peter Little, Michael Horowitz, and A. Nyerges (Boulder, Colo.: Westview Press, 1988), 195–212; Paul Richards, "Ecological Change and the Politics of Land Use," *African Studies Review* 26 (1983): 1–72.
49. Alain de Janvry and Raúl García-Barrios, *Rural Poverty and Environmental Degradation in Latin America: Causes, Effects, and Alternative Solutions* (Rome: Institute for Food and Agricultural Development, 1989).
50. Sometimes this contrast flows from the label *global* environmental issues being used to refer to *international* environmental issues; see note 9.
51. A third path, in which analysts point to the existence of rich and poor groups, peoples, or nations, follows an unstable middle ground between analyses assuming uniform, undifferentiated units and those based on differentiated dynamics. Often the analyst shifts attention to one group on its own, usually the poor, and proceeds with an undifferentiated analysis; see note 45.
52. Nordhaus, "Greenhouse Economics." This economic research is another illustration of proposition 1. In his economic analysis, Nordhaus, like the economists who criticized the LTG, preserves a privileged role for the likes of himself when he steers policymakers against the new advice of planetary scientists and their environmental allies.
53. Anil Agarwal and Sunita Narain, *Towards a Green World: Should Global Environmental Management Be Built on Legal Conventions or Human Rights?* (New Dehli: Centre for Science and the Environment, 1992); and Agarwal and Narain, "Global Warming in an Unequal World." These and allied oppositions to global environmental science, it should be noted, center more on disparities among nations than on the differentiated economic and political conditions within nations—a particular construction in its own right; see the section titled "Sites of 'Deconstruction' of Global Environmental Change," in Taylor and Buttel, "How Do We Know We Have Global Environment Problems?"
54. See also Ronnie Lipschutz and Ken Conca, eds., *The State and Social Power in Global Environmental Politics* (New York: Columbia University Press, 1993).
55. For remarks clarifying the intended scope of my critique, see notes 9, 11, 29, 36, 45, 50, and 51.
56. Pratap Chatterjee and Matthias Finger, *The Earth Brokers* (New York: Routledge, 1994).
57. Boehmer-Christiansen, "A Scientific Agenda for Climate Policy?"
58. Jesse C. Ribot, "Forestry Policy and Charcoal Production in Senegal," *Energy Policy* (May 1993): 559–85, and "From Exclusion to Participation: Turning Senegal's Forestry Policy Around?" *World Development* 23, no. 9 (1995): 1587–99.

59. Judith Carney and Michael Watts, "Manufacturing Dissent: Work, Gender and the Politics of Meaning in a Peasant Society," *Africa* 60, no. 2 (1990): 207–41.
60. Richard A. Schroeder, "Contradictions along the Commodity Road to Environmental Stabilization" (in this volume). See also Richard A. Schroeder, "Shady Practice: Gender and the Political Ecology of Resource Stabilization in Gambian Garden/Orchards," *Economic Geography* 69, no. 4 (1993): 349–65.
61. Nancy Lee Peluso, "Coercing Conservation: The Politics of State Resource Control," *Global Environmental Change* 3, no. 2 (June 1993): 199–217.
62. See note 36.
63. Although such a polemic will not draw everyone over to my side, any GEer objecting to it should notice the popularity of crisis rhetoric in GE discourse (see, e.g., in note 36, "Life itself is at stake"). In Taylor and García-Barrios, "The Dynamics of Socio-Environmental Change," we argue that feeding on fears about the future to promote policies makes coercion and violence become more likely. Coercion is not just an abstract possibility, but one that environmentalists more generally must pay attention to.
64. Taylor, "Re/constructing Socio-Ecologies" and "Building on Construction."
65. Ibid. Although not based in particular cases, a broad contextualization of global environmental science is provided by Buttel and Taylor, "Environmental Sociology and Global Environmental Change."
66. To the extent that novel or contested aspects of the social studies of science are employed in this essay, I am also not dealing with the particularities, messiness, and other difficulties of achieving change in social studies of science. Moreover, this critique connects social studies of science, environmental science, and environmental activism, building in a privileged role for someone like myself whose work spans these three areas.
67. In a fully reflexive extension of the perspective of heterogeneous construction and of proposition 1, I could have teased out the many diverse resources that facilitated, paradoxically, my avoiding the task of such differentiated analysis of knowledge making.
68. See, in addition, the section titled "Sites of 'Deconstruction' of Global Environmental Change" in Taylor and Buttel, "How Do We Know We Have Global Environmental Problems?"
69. See the Afterword to this volume.

Do Androids Pulverize Tiger Bones to Use as Aphrodisiacs?

Simon A. Cole

Her teeth parted and a faint hissing noise came out of her mouth. She didn't answer me. I went out to the kitchenette and got out some Scotch and fizzwater and mixed a couple of highballs. I didn't have anything really exciting to drink, like nitroglycerin or distilled tiger's breath.
▸ *Raymond Chandler,* The Big Sleep *(1939)*

The Tragedy of Extinction

On Uncompahgre and Red Cloud Peaks in the San Juan Mountains of Colorado, the Uncompahgre fritillary butterfly is becoming extinct.[1] The Uncompahgre fritillary never should have been there in the first place. The climate was perfectly habitable ten thousand years ago during the Ice Age, but the butterfly failed to retreat with the glaciers and ended up trapped in the mountains, thousands of miles from its proper arctic climate. Facing several consecutive years of warm weather, the butterfly has steadily climbed the mountain in search of cooler climes. Now it has reached the top, and it can climb no more. "It's being ecologically squeezed off the top."[2] When the end finally comes, say conservation biologists, the Uncompahgre fritillary will be merely one of hundreds of simultaneously occurring extinctions that we happen to notice. In short, this sort of thing happens all the time. The Uncompahgre fritillary serves as a synecdoche for the phenomenon of mass extinction.

The specificity is poignant, the generality tragic, but the question is: why is it tragic *for us*? Increasingly, the answer is being provided and packaged for us by science, represented by a discipline known as conservation biology. What do conservation biologists do? One thing they do is try to prevent extinctions; the other is to chronicle their occurrence. In the words of Hugh Britten, a conservation biologist at the Nevada Biodiversity Research Center who studies the Uncompahgre fritillary, "I am presiding over the extinction of this species."[3] Just as priests or shamans preside over ceremonies of passing in other cultures, conservation biologists document and interpret the passings of species for Western scientistic culture. The problem is that these passings are so frequent that science can only hint at their magnitude. Conservation biologists, by attempting to quantify the rate of extinction, by drawing the public's attention to the imminent extinction of charismatic megafauna like the tiger, panda, rhinoceros, whale, or elephant, render the tragedy of extinction palpable to their reading and viewing audiences. But although the sense of sorrow, and even shame, that we feel when confronted with the spectacle of extinction is very real, it has done little to stem the tide of extinctions. Today, conservation biologists are becoming increasingly convinced that any positive change in our relationship with other species demands something more powerful than pity.

Conservation biologists are now beginning to appeal to our own economic self-interest in a finite chemical resource. "The loss of any species should be considered a tragedy," says Edward O. Wilson, because "every organism—animal, plant, microorganism—contains a million to ten billion bits of information in its genetic code, hammered into existence by an astronomical number of mutations and episodes in natural selection."[4]

But conceiving species as information capital seems a rather crass justification for preserving them, as some conservation biologists, who see economic arguments as unnecessary concessions to a materialist ethic, readily admit.[5] A second reason for the tragedy is based on feeling rather than reason. We are somehow moved by the slow death of the Uncompahgre fritillary. But why? At bottom, extinction is merely the death of an individual, a common enough occurrence in a brutal world, but something in an extinction compels a stronger response, akin to empathy.

It is the *last* butterfly, like George Schaller's "last panda,"[6] that provokes an empathic response in its human observers. It is the idea of being the last of one's kind that we find so disconcerting. This evokes a loneliness we would not wish on ourselves, as we are reminded when we read the story of Ishi, "the last wild Indian in North America."[7] Ishi lived his whole life as a member of a dwindling band of Yahi trying to survive the encroachment of white settler society by retreating further and further up the slope of Mount Lassen, in much the same manner as the Uncompahgre fritillary. Following the death of his mother, Ishi spent an unknown period of time, possibly as long as three years, alone. Although unaware that he was "the last wild Indian," Ishi must have understood in some way that he was the last Yahi.

Although Ishi was neither the last of his "race" nor of his species, his story anthropomorphizes the tragedy of extinction. In those last years in the wild, his situation was analogous to that of the last Uncompahgre fritillary. Alone, trapped, pursued by climates or cultures that they only vaguely understand, neither can find a way of going on. By "going on," I mean both continuing to struggle as an individual and reproducing, an alternative means of going on. For the Yahi who tried to survive in hiding,

> those who remained were hopelessly crippled not solely because they had suffered the loss of two thirds of their number, but because amongst those two thirds were almost all their young. The real hazard to the possible success of the long concealment may have been that those who were left faced a future in which they shared no sure investment.[8]

After a few years alone, Ishi could not go on. It was at this moment that he wandered out of the mountains and into nonnative America.

> Ishi's arrival at the slaughter house was the culmination of unprecedented behavior on his part. A few days earlier, without hope, indifferent whether he lived or died, he had started on an aimless trek in a more or less southerly direction which took him into a country he did not know. Exhaustion was added to grief and loneliness. He lay down in the corral because he could go no farther. He was then about forty miles from home, a man without living kin or friends, a man who had probably never been beyond the borders of his own tribal territory.[9]

Although Ishi had every reason to anticipate murder—and, indeed, his feet carried him, of all places, to a slaughterhouse—it turns out that he found a way, albeit unconventional, of going on. The anthropologists who took charge of him cataloged his language, collected his artifacts, and, most important, recorded his story. In narrative, Ishi found a form of immortality. His genes did not go on, but his story, or some version of his story, did. There might have been, of course, other endings to the story. Ishi might have died alone in the mountains; he might have dropped out of history instead of finding himself a place in it. Ishi might have reproduced following his rescue. Or he might have been captured and sold into intermarriage, as some of his female cousins apparently were. (And, in fact, it is they, not Ishi, who have dropped out of history.) Some Indian tribes were offered the opportunity to assimilate; others, including the Yahi, were not.

The Uncompahgre fritillary may also find a way of going on. When pressed, biologists become less confident about predicting the imminent extinction of the Uncompahgre fritillary. Population fluctuations are difficult to interpret, population counts are unreliable, and—the most tantalizing possibility—additional colonies may yet remain undiscovered.

> Even Dr. Britten, who has made numerous searches, admits it is impossible to be sure whether or not there are hidden colonies in the wilderness of the San Juans. It is particularly difficult because the butterflies are visible and in

> flight for only about three weeks a year, in July. "There have been reports of additional colonies by one other lepidopterist who is refusing to reveal where they are."[10]

The Palos Verdes blue butterfly, long presumed extinct, turned up in a meadow in southern California in 1994. The Uncompahgre fritillary may yet have some tricks up its sleeve.

Do Androids Dream?

This essay inquires into our responses, empathic and opportunistic, emotional and economic, to the plight of "others" struggling to find a way to go on. It is difficult for anyone, scientist or citizen, to make sense of a world where species are being extinguished faster than we can count them, a world in which biologists perform "triage" on endangered species,[11] Latin American nations are praised for selling their rain forests to U.S. pharmaceutical companies, buffalo burgers are environmentally correct, and zoos are no longer prisons for animals but safe havens.

If we cannot make sense of what is happening, we can at least try to understand the ironies and ambiguities that face us. This essay is based on the rather fanciful idea that the fertile imagination of the late science-fiction writer Philip K. Dick and his landmark 1968 novel *Do Androids Dream of Electric Sheep?* may be useful guides. Dick's novel is best known by Ridley Scott's film adaptation, *Blade Runner*. Thanks to its retro-futuristic images of postapocalyptic Los Angeles, gritty film noir atmosphere, and compelling treatment of issues dear to postmodern culture, such as the erasure of the human/nonhuman divide, *Blade Runner* has generated both a cult following and a copious body of scholarly literature.[12] The film itself is a contested text with versions of differing degrees of authenticity. The screenplay went through numerous controversial revisions, and devotees fiercely debate whether Dick himself approved of the final version before his death, just before completion of the film in 1982.[13] Scott's original cut of the film was released briefly and did poorly at the box office, so the studio released a shorter, more comprehensible, narrated version. In 1992, Warner Brothers rereleased "The Director's Cut," which true fans regard as the more authentic (and more interesting) work.

The title of Dick's novel asks what makes us human. Do androids dream? If they do, if they have emotional lives, humans will be hard-pressed to maintain the boundary between themselves and the cyborg "other." This is the issue that has commanded the attention of most treatments of the book and the film, but in fact Dick asks not whether androids dream, but "Do androids dream of *electric sheep*?" The animal other is crucial to Dick's exploration of what it means to be human. The animal theme was largely omitted in the film version; it shows up only subtly when at all, and most critics have ignored it.[14] By bringing out the role that animals play in the original text, I hope to show the relevance of Dick's imagined moral dilemmas to our own attempts to come to terms with our place in a world that seems to be hurtling toward its own self-destruction.

Do Androids Dream of Electric Sheep?

Do Androids? tells the story of Rick Deckard, a policeman (called a "blade runner" in the film) whose job is to "retire" any androids that manage to escape to earth from their enslavement in the "off-world" colonies. Androids are physically indistinguishable from humans, so the police identify them by testing them with a polygraph-like apparatus that measures their emotional response to a series of provocative questions. The paradox is that in order to continue to function effectively as an assassin and interrogator, Dick's noir hero must suppress his emotions to the point that his artificial targets' emotional lives appear richer than his own.

The novel chronicles Deckard's odyssey through the techno-philosophical problem of distinguishing real humans from fakes, continuing the inquiry of his namesake, René Descartes, who puzzled about the

> instance of human beings passing by in the street below, as observed from a window. In this case I do not fail to say that I see the men themselves, just as I say that I see the wax; and yet what do I see from the window beyond hats and cloaks that might cover *artificial machines, whose motions might be determined by springs*? But I judge that there are *human beings* from these appearances, and thus I comprehend, by the faculty of judgment alone which is in the mind, what I believed I saw with my eyes.[15]

It is the exploration of what distinguishes humans from androids that makes the film so compelling. But what is not explained in the film is that the emotional responses tested by Deckard's apparatus are all provoked by scenarios involving animal suffering. Why *animal* suffering? Because "animal empathy" is the one aspect of humanity that androids are unable to fake. In the future society imagined by the novel, which I will call, for lack of a better term, "blade-runner society," animal empathy is the highest virtue.[16] The historical explanation for this peculiar social value lies in the mass extinction of most animal species because of environmental degradation following "World War Terminus." The remaining animals are protected by strict laws and held as spiritual totems. Directly following the war, all citizens were required to care for an animal of some kind. Caring for an animal is enforced not by law but by social pressure: lacking a pet is viewed as an ethical and social lapse. Pets have replaced automobiles as status symbols. Neighbors vie to outdo one another by possessing rarer, costlier animals. In a society where everyone loves and covets animals, androids are exposed by their lack of animal empathy. Androids, it seems, do *not* dream of electric sheep, and that is their undoing when a blade runner catches up with them.

Do Androids Dream of *Electric* Sheep?

But we may read Dick's title another way. Do androids dream of *electric* sheep? With animals so rare and yet so highly prized as status symbols, a flourishing market in artificial pets has arisen. Deckard, in fact, can only afford an electric sheep on his civil service salary, but he is tormented by the inadequacy of his bogus sheep and obsessed by his desire for what he calls a "real animal."[17] Deckard's longing for an animal companion is at once mercenary and spiritual. In the same breath, he articulates his spiritual need to care for a live animal and calculates the number of bounties for android retirements he would need to be able to afford it. Moreover, Deckard is consumed by jealousy of his neighbors' "real" pets. Although Deckard earnestly wants to nurture an animal, he cannot shake the nagging feeling that he also wants a real animal for the social prestige that it carries.[18]

In blade-runner society, animals have become both status symbols and objects of genuine love, and, although sometimes themselves of questionable authenticity, they have become the wedge with which the "real" is distinguished from the "fake" among humanoids. Whereas artificial animals' ability to "pass" is viewed as a necessary social lubricant, androids' even greater ability to "pass" is dangerous. Androids that attempt to pass on earth must be "sniffed out"—by emotionally deadened humans and by animals, most of which are mere representations. It may be true that, as Donna Haraway says, "The cyborg appears in myth precisely where the boundary between human and animal is transgressed,"[19] but it would also appear that animals police the boundary between humans and cyborgs, extending the role that animals already play in policing: we now employ dogs (and pigs) to sniff out truth from falsity, legitimate cargo from contraband. And it is in the dystopian future posited by a science-fiction film contemporary with *Blade Runner*, James Cameron's *Terminator* (1984), that dogs are employed to sniff out cyborg infiltrators because they, unlike humans, are capable of distinguishing fake humanoids from the real thing.[20]

This situation becomes even stranger in the case of Phil Resch, a fellow bounty hunter whom Deckard encounters. In a plot twist too complex to have been incorporated wholesale into the film, Resch is told that he is an android himself. Resch is deeply shaken by this revelation, as one might well imagine. But what he finds most difficult to comprehend is his relationship with his animal. Resch protests: "I own an animal; not a false one but the real thing. A squirrel. I love the squirrel, Deckard; every goddamn morning I feed it and change its papers—you know, clean up its cage—and then in the evening when I get off work I let it loose in my [apartment] and it runs all over the place."[21]

Resch's remonstration disturbs Deckard, who considers himself a "real" human but cares for an electric sheep.

At another point in the book, the Rosen Association (called "the Tyrell Corporation" in the film), manufacturer of androids, tries to bribe Deckard, whose weakness they easily discern, with an owl, a supposedly extinct animal. Is the owl really an illegally obtained rare animal, affordable only to large corporations, as Rosen claims? Or is it merely an elaborate fake?

This owl is one of few animal symbols preserved in the film.[22] It appears in a scene with the android Rachael, to whom the screenwriters have assigned Resch's dilemma—she is an android, but she does not know it yet.

"Do you like our owl?" Rachael asks.

"Is it expensive?" Deckard replies.

"Very."

"It's fake, isn't it?"

"Of course it is . . . I'm Rachael."

The situation becomes all the more interesting since Rachael and Deckard end up sleeping together. In the book, a jilted Rachael punishes Deckard by pushing his *real* sheep, purchased with bounties from retiring her fellow androids, off his roof. Rachael's anger at Deckard's inability to love her—and, worse, his insistence on using his affection for a sheep to draw distinctions between them—suggests a love triangle, or at least an "empathy triangle," between humans, androids, and animals.

Another animal image appears in a dream sequence omitted from the commercial release of the film (but reinserted in the "Director's Cut") in which Deckard sees a galloping unicorn. The scene is important not only for introducing the idea of extinction, symbolized by the mythical unicorn, but also as crucial evidence for those aficionados who adhere to the theory that Deckard is himself an android.[23] Whether he is an android or not, Deckard is so divorced from the natural world that most animals exist only in memory or myth. We are in danger of sharing Deckard's isolation. As *Time* warns in a report on the extinction crisis, "all too soon, dreams may be the only place where tigers roam freely."[24]

Episodes in Extinction

Today, responses to the endangered species crisis are turning increasingly toward economic incentives to preserve wildlife. Efforts are now being focused on uniting economic and ecological goals—"making conservation pay." Such efforts take many forms: ecotourism, wildlife ranches in Africa, debt-for-nature swaps, captive breeding, and biodiversity prospecting. In all these cases, the aim is generally the same: to convince some reluctant, usually poor, nation or region that allowing extinction to occur is simply poor resource management. Timber and cattle may appear profitable in the short run, conservationists argue, but in the long run, species diversity will be more profitable as a sustainable resource, whether as spectacle for tourists, nuts for Ben & Jerry's Rainforest Crunch, ingredients for skin lotion, quarry for big-game hunters, or raw material for pharmaceutical firms. Consider, for example, the following vignettes from the strange world of endangered species preservation—not a fictional world, but our own:

- The Instituto Nacional de Biodiversidad (INBio), founded by conservation biologist Daniel Janzen and the Costa Rican government, is based on the premise that biodiversity is best preserved by commercializing it. In 1991, INBio signed an agreement with the pharmaceutical giant Merck, selling the rights to useful products emerging from INBio's project of locating and cataloging the species of Costa Rica's exceptionally rich biota. The Merck-INBio deal has been almost universally praised in conservation circles as a "win-win" agreement.[25]
- "In Zimbabwe, to promote the conservation of the wildlife resources found on communal lands, private game reserves have been established where revenues from hunting are paid to local communities. Recreational hunting is now the most positive and widespread economic incentive for the conservation of large mammals in Zimbabwe."[26]
- Meanwhile, in China, black-market entrepreneurs are reportedly breeding tigers in captivity to supply the herbal medicine market with pulverized bones and other parts. "A tiger-breeding farm in northeast China that started with 14 animals in 1986 now has 62 Siberian tigers. With modern techniques, it will be possible to breed 2,000 'industrial' tigers every seven

years."[27] Since poachers have decimated the wild tiger population, commercial captive breeding of tigers appears to be smart resource management. It just might also save the tiger from extinction.

- Antonie Blackler, a geneticist, is experimenting with a biotechnological conservation method. He is trying to impregnate common frog species with embryos from endangered species, thus enabling common animals to serve as surrogate mothers for rare ones. In theory, he argues, the same method may be applicable to large mammals.[28]
- Arguing that habitat preservation on a large scale is neither politically nor technically feasible, some conservation biologists, such as Michael Soulé, are increasingly turning to biotechnological methods for preserving endangered species. Since "biotechnology is accelerating at a pace that could not have been foreseen thirty years ago," it promises far greater rewards in the future than low-tech methods like habitat preservation and conventional captive breeding.[29] Among the methods Soulé expects to flourish in the twenty-first century are cryogenics, DNA "fingerprinting," cloning, gene transplants, and automated taxonomy. He suggests that it might be possible to bank gametes of all vertebrates, but he concedes, "Some biologists might object to the idea of 'cryoconservation' on ethical grounds."[30] Indeed, Dale Jamieson asks, "In doing this, aren't we using animals as mere vehicles for their genes?"[31] Cryopreservation, unlike habitat preservation, treats extinction solely as a reproductive problem and endorses a concept of extinction that includes neither the animal's culture—its "wildness"—nor its role as a member of an interdependent ecosystem, but only its genetic code.

In short, the conservation community is striving by other means to attain the same goal sought by blade-runner society: the merger of avarice and sentiment into a single force for the preservation of animals. The valorization of animals is achieved by reducing them to their constituent parts. Minute parts of tigers' bodies, their glands and genitals, are now even more valuable than big parts, like their pelts. Biodiversity prospecting takes this reductionism to its fullest extent; it posits an economic system in which the value of an animal is located in its chemicals. In the words of Thomas Eisner, the guru of biodiversity prospecting:

> Although we are beginning to grasp that extinction is forever, we have yet to comprehend what we lose when species disappear. The point that cannot be overemphasized is that biotic impoverishment is tantamount to chemical impoverishment. Loss of a species means a loss of chemicals that are poten-

> tially unique in nature, not likely to be invented independently in the laboratory, and of possible use. Aside from other measures of worth, species have chemical value.[32]

A nonentity fifty years ago, DNA is now being touted as *the* natural resource of the twenty-first century.

The difficult choices raised by this new genetic reductionism are illustrated by a bizarre episode in the effort to preserve the endangered orangutan. Until very recently, conservationists, acting on the general principle that genetic diversity strengthens the gene pool, interbred Bornean and Sumatran orangutan, assuming that they were different strains of the same species. But genetic analysis of the animals revealed that the two strains of orangutan "are more genetically different from one another than lions are from tigers."[33] A benign intervention suddenly became an unwarranted act of "pollution," the destruction of a "pure" species through interbreeding, and zoos are now sterilizing orangutans of mixed parentage. What we should notice about this "ape version of racism"—besides the supersession of a morphological taxonomy, based on physical appearance and structure, by a reductionist genetic taxonomy—is the ascendance of a new system for valuing animals, according to their genetic purity.[34]

Among the primary tasks of the new breed of conservation biologist is the construction of a market for genetic materials. As David Takacs observes, the conservation biologists and parataxonomists at INBio "*love* biodiversity. It is their life's blood. But to sustain this love, they need to sell off the objects of their affection, and fast."[35]

These conservation efforts provoke the same uneasy feeling that American carnivores experience when they are told that they can help support the resurgence of the bison by eating buffalo burgers, instead of South American beef implicated in rain-forest destruction. There is something inherently disturbing about devouring what you wish to preserve. Meanwhile, as habitat destruction accelerates, conservationists are being subjected to increasingly excruciating tests of how far they will go in order to preserve endangered species. Peter Jackson, chairman of the Cat Specialist Group of the World Conservation Union, says he is "tortured" by the prospect of industrial breeding but that it must, nonetheless, be considered one of the only remaining opportunities to protect tigers from extinction.[36] It is a grim irony that the tiger's last refuge might be industrial breeding farms that serve the very market that drove it from the wild.

Some of these conservation initiatives certainly do appear to embrace the very values that led to endangerment in the first place. Private game farms "reserve" large mammals for use in latter-day "great white hunts," the very same unsustainable hunting practices, according to most ecologists, that were responsible for endangering the animals in the first place. These farms continue the European elites' practice of reserving African mammals for their own pursuit, pleasure, and prestige.[37] The private game reserve program in Zimbabwe, for example, is designed to be profitable for ranchers in the long run, but for now "the social incentive of the prestige of having a black rhino on their land has been sufficient to encourage a number of ranchers to apply for such responsibilities."[38] Historians have suggested that this elitism lies at the root of the species preservation movement in Africa.[39] Similarly, critics have argued that INBio, by commodifying biodiversity, perpetuates the same values that caused the extinction crisis in the first place.[40] But despite these qualms, the seemingly odd dynamic of using consumption as a form of preservation is beginning to take on an air of normalcy, or even, in some cases, a moral urgency.

Do Electric Humans Dream of Sheep?

Dick's suggestion that our ability to empathize with animals makes us human beings instead of machines is echoed in contemporary culture by Edward O. Wilson, who has even coined a term for the feeling of love of nature. For Wilson, "biophilia" is an evolved human genetic trait dictated by the inexorable logic of natural selection. Wilson has thus conceived a new method of linking selfishness and sentiment in order to facilitate conservation. Biophilia yields evolutionary, rather than economic, gain. The "selfish gene" is an even more powerful and self-interested engine than avarice.

> For if the whole process of our life is directed toward preserving our species and personal genes, preparing for future generations is an expression of the highest morality of which human beings are capable. It follows that the destruction of the natural world in which the brain was assembled over millions of years is a risky step. And the worst gamble of all is to let species slip into extinction wholesale, for even if the natural environment is conceded more ground later, it can never be reconstituted in its original diversity.[41]

Besides Wilson's appeal to self-interest, we cannot help but notice the persistent reproductive undertone to discourse about species extinction, the recurring association of *human* sterility with an impoverished natural landscape.[42] In *Do Androids Dream of Electric Sheep?* the damage wrought by nuclear radiation in "World War Terminus" is not restricted to animals.

Human fertility has diminished as well, and radiation has left many mentally and/or physically incapacitated humans, called "specials." Humans whose reproductive capacity remains intact are encouraged to emigrate "off-world": "The U.N. had made it easy to emigrate, difficult if not impossible to stay. Loitering on Earth potentially meant finding oneself abruptly classed as biologically unacceptable, a menace to the pristine heredity of the race. Once pegged as special, a citizen, even if accepting sterilization, dropped out of history. He ceased in effect to be part of mankind."[43]

The "specials" thus find themselves in much the same position as members of endangered species: they are the last of their kind, destined to "drop out of history," doomed by their legally dictated inability to reproduce in sufficient numbers.

There is, therefore, an air of denial surrounding the outpouring of concern for animal welfare in blade-runner society. It is not clear that humans, faced with the choice between extinction or self-imposed exile, are in any position to pity other creatures. The tragedy that humans project onto animals is their own, and animal *empathy* is just that.

Do Imperialists Dream of Electric Natives?

Today as well, animal extinctions are metaphors for our own reproductive anxiety, and for concerns about human extinction and genetic purity. The rhetoric of biodiversity does not distinguish between the human and the nonhuman. Calls for cataloging the genes of vanishing species are accompanied by calls for preserving the genes of vanishing races.[44] Like the endangered species of the rain forest, the postmortem on the endangered "races" of the rain forest has already been performed by the appropriate scientific institutions. With impending extinction presumed, the Human Genome Diversity Project (HGDP) proposes to cryogenically preserve indigenous peoples' DNA before it is too late. It is assumed that other means of reproduction of this precious genetic material, such as miscegenation, will not occur. This racist assumption actually accompanies the valorization of these same peoples' genetic material. First Tier science is at last prepared to hybridize with the indigene, but only on its own terms, through mediations of ritual purity and prophylaxis:[45] the freezer, the syringe, the polymerase chain reaction. Science offers to preserve racial purity and a valuable natural resource simultaneously.

Biodiversity, human and nonhuman, serves as a potential resource for genetic engineering—of drugs, agricultural products, or, indeed, human beings. Whether the object of study is human or not, genetic surveys inevitably undergo successive phases of knowledge and exploitation: knowledge facilitates exploitation. "All of the information—ecological, chemical, behavioral, genetic, etc.—to be gathered on Costa Rica's biodiversity can be organized, cross-referenced, manipulated, and offered to the country, region, and world through the public domain and commercial sales," INBio's managers tell us.[46]

The HGDP represents the ultimate manifestation of anthropology's imperialist project. No longer content with recording the ritual structure of "primitive" people like Ishi, First Tier scientists now wish to extract and catalog their genetic structure as well. Genetic surveys prepare the way for the First Tier self to plunder genetic resources in order to reconstruct itself. "Anthropologists of possible selves," writes Haraway, "we are technicians of realizable futures."[47]

In this territory too we find our way mapped by science fiction, in this case Octavia Butler's novel *Dawn* (1987). In *Dawn*, the imperialists are extraterrestrials, the Oankali. As Haraway puts it, "their own origins lost to them through an infinitely long series of mergings and exchanges reaching deep into time and space, the Oankali *are* gene traders." Like our First Tier scientists, the Oankali offer preservation, of a sort, to a doomed people, in this case the entire human race. But, of course, the form of preservation they offer—merger and gene exchange—carries a price: the loss of identity, of the distinguishable, pure self. Once again, extinction engenders commerce: the closer a species or race approaches extinction, the more it awakens mercantile interest in a gene-hungry universe. Endangered species, indigenous people, and First Tier beneficiaries of genetic engineering might say, with the Oankali, that "their essence is embodied commerce."[48] Given this close relationship between extinction and trade, it is hardly surprising that the punishment meted out to endangered species traders was to impose trade restrictions on Taiwan and threaten China with them.[49]

Do Chinese Dream of Electric Tigers?

Ecologists tend to blame most environmental degradation, such as rain-forest destruction, desertification, global warming, ozone depletion, and habitat destruction, on the consumption patterns of industrialized countries. These countries' appetite for timber and beef, for instance, creates economic incentives for less-developed countries to degrade their own native ecology in order to supply these resources.

In the case of endangered species, however, the moral ground shifts. Although it is again Third Tier peasants who carry out the actual destruction, this time the offending appetite is located not in the industrialized world, but in the "tradition-bound" consumer nations of Asia: China, Taiwan, and Korea. The "insatiable demands of the Asian medical market" for rhinoceros horns, tiger bones, and other exotic wildlife products is driving these animals to extinction.[50] The *New York Times* describes the crisis as follows:

> The trade is driven by booming markets for ancient Chinese medicines and potions made from tiger parts. In Hong Kong, China, and Taiwan, and in Chinatowns across Europe and North America, Chinese apothecaries do a steady trade in tiger wines, tiger balms, and tiger pills, celebrated among Chinese and other Asian peoples for their supposed powers to treat rheumatism, to restore failing energy and to enhance sexual prowess, as well [as] for the treatment of rat bites, typhoid fever, and dysentery, among other ailments.[51]

The problem is presented simply as one of Asian consumption, ignoring a long history of *global* commerce that has driven not only trade in tigers and tiger parts, but also, far more significantly, the widespread destruction of the tiger's habitat. (It should also be noted that Chinese herbal healers are, within a different medical regime, exploiting tigers for precisely the same purpose Merck is exploiting the Costa Rican rain forest.)

With Westerners valuing nature for sentimental reasons, and Asians treating animals as commodities, stereotypic roles have been reversed. Asian culture is supposedly more attuned to living in harmony with nature, in contrast to the European tendency to exploit nature and ravage landscapes.[52] Now Westerners are the spiritual sentimentalists, while Asians become the relentlessly rational economic actors they have long been criticized for *not* being.[53]

After decades of "great white hunting" in Africa, after the great buffalo slaughter of the American West, Westerners have at last found remorse and conservation, only to find their best efforts stymied by a new generation of hunters, traders, and wasteful, profligate consumers. Our frustration has primarily been vented by criticizing Asian "values," which allow Asians to remain indifferent to the ultimate fate of the species. Western ecologists employ the same technique as Deckard, a test of animal empathy, to distinguish themselves from an inhuman other.[54]

We could save these magnificent beasts, we lament, if we could only wean those Asians from their superstitious beliefs. Surely someone has already thought of peddling ersatz pulverized tiger, but flooding the market with knockoffs would only stimulate demand for the real thing.[55] Amid a farrago of imitations and fakes, we are again reminded of their curious tendency to usurp and yet still further valorize the "real."

The criticism can occasionally lapse into castigating Asians simply for being so darn numerous: "When advances in hunting techniques are combined with lower trade barriers and rapidly growing populations that demand medicines made from exotic wildlife, an entire species can be wiped out in one generation."[56]

This reference to Asian overpopulation is hardly accidental, especially in light of the prominence given to the sexual angle of Asian demand for tiger parts. The American press has pinned a large portion of the blame for the tiger's decline on its use as an aphrodisiac by superstitious cultures, just as it did with the conservation movement's last "poster" species, the rhinoceros, whose horn is also valued for its aphrodisiac properties: "Affluent Taiwanese with flagging libidos pay as much as $320 for a bowl of tiger-penis soup, thinking the soup will make them like tigers, which can copulate several times an hour when females are in heat."[57]

The portrayal of the tiger's extinction as the simple effect of the inadequate, or depraved, sexuality of an excessively wealthy emerging Asian elite belies the tragic complexities and ambiguities of extinction. Moreover, the tiger's situation reveals the hazards of the free-market approach to conservation: we must valorize endangered species in order to preserve them, but not too much. In an unstable social ecology, highly prized species can be quickly driven to the brink of extinction. It will be extremely difficult, as well as morally ambiguous, for conservationists to try to distinguish between the "right" kind of valorization and the "wrong" kind, between exploitation and the new "profitable" conservation.

The controversy over tigers is only a harbinger of the global struggle for control of the earth's most recently discovered natural resource: its genes. The location of most of the world's uncataloged biodiversity in capital-poor regions will only heighten wealthy nations' ardor in the race to secure access to these resources in the coming gene wars.

Do Androids Pulverize Tiger Bones to Use as Aphrodisiacs?

Do androids pulverize tiger bones to use as aphrodisiacs? They would, but not because they are stupid, superstitious, cruel, or unempathic. They would because, like any other living thing, they will do what they have to do in order to go on. This common urge to go on is what unites animals, humans, and possibly—only the future will tell—androids. How might androids go on? As Haraway argues, androids will be compelled to devise new and innovative solutions to the problem of going on. Might they try to manufacture new and better-living machines themselves, literally a form of (re)production?[58] They will try that and more, perhaps even eating pulverized tiger parts.

The question about androids, then, is: do they struggle to go on? An android with the desire to find a way to go on is morally and practically indistinguishable from a human being; an android that cannot find a way, that "dies" at its appointed (by its maker) hour, is just a machine. The androids in Dick's original text are of the latter type. When Deckard threatens to kill Rachael,

> the dark fire waned; the life force oozed out of her, as he had so often witnessed before with other androids. The classic resignation. Mechanical, intellectual acceptance of that which a genuine organism—with two billion years of the pressure to live and evolve hagriding it—could never have reconciled itself to. "I can't stand the way you androids give up," he said savagely.[59]

The androids in *Blade Runner* are not like those in the book. The most brilliant coup of the screenplay lay in making Roy Baty, leader of the renegade androids, the "king of the constructed," to some extent the hero of the movie. The dramatic force in the film lies not with the deadened executioner Deckard, but with Roy in his search for his father and maker. It is Deckard himself who observes that Roy possesses all the trapping of the legendary dramatic hero of uncertain paternity: "All he wanted were the same answers the rest of us want. Where do I come from? Where am I going? How long have I got?"

The dramatic structure of the film is centered around Roy's all-too-human efforts to go on. He mates with another android, but to no avail. He breaks into the Tyrell Corporation's headquarters, where he demands repairs to the fail-safe system that restricts his genetically engineered body to a four-year life span. Neither polite persuasion nor savage threats are of any use. Finally, struggling to stay alive just long enough to have his revenge, he desists from killing Deckard at the last moment. Deckard muses, "maybe in those last moments he loved life more than he ever had before. Not just his life: anybody's life. My life." It is empathy that inspires Roy to eschew vengeance and let Deckard live. Like Ishi, the last of his kind, Roy achieves immortality of a sort by storytelling. He passes some version of his story, no matter how brief and incomplete, on to Deckard, his enemy and the murderer of his people, much as Ishi was forced to tell his story to white social scientists. "I've seen things you *people* wouldn't believe," Roy says contemptuously in his final testimony. "Attack ships on fire off the shoulder of Orion. I watched sea beams glitter in the dark near the Tannhauser gate. All those . . . moments . . . will be lost in time . . . like . . . tears . . . in rain. Time to die."

All the tension and boundary drawing between humans, animals, and androids, then, can be ascribed to a struggle for inclusion in a common reproductive project. The criteria for inclusion are not consistent, nor are the winners selected according to neat distinctions between humans and nonhumans. Instead, genetic engineering, like all forms of (re)production, is shaped by elements of both love and exploitation. Some participants, like the First Tier and the Oankali, will be in the game by virtue of their superior strength. Others may choose to join in as the most palatable means of escape from a difficult situation. Others, like the "specials," may be excluded altogether. And, of course, dark horses and Trojan horses, of which we may not even be aware, will be involved. We are all hosts to parasites and parasites within parasites.[60] Lynn Margulis and Dorion Sagan, for instance, speculate that space travel allows humans to function as vehicles for microbes contained within our bodies. Although humans might "go extinct" in the conventional sense, we may well gain immortality for our role in facilitating a galactic "microbial diaspora."[61] While opportunistic microbes are winging their way across space, cryogenic gene banks, perhaps deep underground, when their long-lasting power supplies give out, might simply repeat the extinction of the life-forms they "preserved"—the first time a tragedy, the second a farce. Extinction may not be the inescapable destiny that it might at first glance appear to be, for the "other" or for "us." Given imagination and opportunity, there are ways of going on, in some form or another.

Notes

1. An earlier version of this essay appeared in *Social Text* 42 (spring 1995). I wish to thank the Social Text Collective for their assistance. I am also indebted to Peter J. Taylor, Laura Kelly, David Takacs, Paul N. Edwards, Saul E. Halfon, Davydd Greenwood, Beth Drexler, Larry Carbone, Rosaleen Love, Jo Liska, George Kolias, Yoo-Shin Kim, Mary Lui, Clark Troy, Douglas G. Kelly, and the members of the Social Analysis of Ecological Change seminar at Cornell University (spring 1994): Laura Fitton, Gonzalo Kmaid, Hanah LeBarre, Govindan Parayil, Zed Rothman, Theresa Selfa, Sean Selinger, Chris Shields, and Christel Van Arsdale. The ideas expressed in this essay were initially explored in a video that Kavita Philip and I produced (see note 22). We wish to thank Antonie Blackler and Michael Fortun for their invaluable cooperation.

2. Carol Kaesuk Yoon, "Rare Butterfly Consigned to Extinction," *New York Times*, April 26, 1994.
3. Ibid.
4. Edward O. Wilson, "Biodiversity, Prosperity, and Value," in *Ecology, Economics, Ethics: The Broken Circle*, ed. F. Herbert Bormann and Stephen R. Kellert (New Haven: Yale University Press, 1991), 9.
5. David Ehrenfeld, "Thirty Million Cheers for Diversity," *New Scientist* 110 (June 12, 1986): 38–43.
6. George B. Schaller, *The Last Panda* (Chicago: University of Chicago Press, 1993).
7. Theodora Kroeber, *Ishi in Two Worlds: A Biography of the Last Wild Indian in North America* (Berkeley: University of California Press, 1961).
8. Ibid., 10.
9. Ibid., 93.
10. Yoon, "Rare Butterfly Consigned to Extinction."
11. Ibid.
12. See, for example, Peter Fitting, "The Lessons of Cyberpunk," in *Technoculture*, ed. Constance Penley and Andrew Ross (Minneapolis: University of Minnesota Press, 1991), 295–315; David Harvey, *The Condition of Postmodernity: An Enquiry into the Origins of Cultural Change* (Oxford: Basil Blackwell, 1989); and Paul N. Edwards, "The Terminator Meets Commander Data: Cyborg Identity in the New World Dis/Order" (this volume).
13. *Retrofitting* Blade Runner: *Issues in Ridley Scott's* Blade Runner *and Philip K. Dick's* Do Androids Dream of Electric Sheep? ed. Judith B. Kerman (Bowling Green, Ohio: Bowling Green State University Press, 1991).
14. The only exceptions that I found are Marleen Barr, "Metahuman 'Kipple'; or, Do Male Movie Makers Dream of Electric Women? Speciesism and Sexism in *Blade Runner*," in ibid., 25–31; and Norman Fischer, "*Blade Runner* and *Do Androids Dream of Electric Sheep?* An Ecological Critique of Human-Centered Value Systems," *Canadian Journal of Political and Social Theory* 13 (1989): 102–13.
15. René Descartes, *Meditations on First Philosophy* (1641), trans. John Veitch (1901), Meditation II, paragraph 13; emphasis added. I am indebted to Clark Troy for drawing this connection.
16. The theme of empathy is taken to its fullest in Octavia Butler's latest novel, *Parable of the Sower* (New York: Four Walls Eight Windows, 1993). Butler's heroine suffers from "hyperempathy," which causes her to experience directly the pain of people, and some animals, around her. I will discuss Butler's work later in this essay. Douglas G. Kelly has pointed out to me that the earliest exploration of empathy in science fiction may have been Olaf Stapledon's *Star Maker* (1937), in which higher levels of civilization are indicated by the breadth of empathy felt for other forms of existence.
17. Avital Ronell, *The Telephone Book: Technology, Schizophrenia, Electric Speech* (Lincoln: University of Nebraska Press, 1989), points out that the inventor of *electric speech*, Alexander Graham Bell, was also interested in the genetic engineering of *sheep* (337–40). Ronell notes (453) that Benjamin Franklin was also interested in electric sheep, in that he recorded experiments on the subject of the rate of putrefaction of sheep killed by electricity.
18. If this all sounds like science fiction, consider the following excerpt from John Barnard, *The Handy Boy's Book* (London: Ward Lock, n.d.), an early twentieth-century primer aimed at young boys: "Every boy ought to keep at least one pet, but not unless he is prepared to give all of the care and attention necessary to keep it in health and comfort. If you have a real affection for your pet, you will never neglect it; if you have not that affection, you have no right to keep the animal" (238). Here we see the source of

both the social obligation to keep a pet and the anxiety that one is not worthy to do so. On "boy" literature, see Mark Seltzer, *Bodies and Machines* (New York: Routledge, 1992).

19. Donna J. Haraway, "A Manifesto for Cyborgs: Science, Technology, and Socialist Feminism in the 1980s," *Socialist Review* 15, no. 2 (1985): 68.
20. On *Terminator*, see Edwards, "The Terminator Meets Commander Data," in this volume.
21. Philip K. Dick, *Do Androids Dream of Electric Sheep?* (New York: Ballantine, 1968), 112.
22. For a visual explication of the animal imagery in the film, placed in the context of contemporary endangered species conservation initiatives, see the homemade video by Kavita Philip and myself titled *Blade Runner: The Nature Lover's Cut* (1994).
23. In the final scene of the film, Deckard finds an origami unicorn left at his apartment by another blade runner. Since no one else knew about Deckard's dream, the theory is that the dream must have been implanted. Therefore, Deckard, like Rachael, is an android so advanced that he does not even know it.
24. Eugene Linden, "Tigers on Trial," *Time*, March 28, 1994, 44.
25. Elissa Blum, "Making Biodiversity Conservation Profitable: A Case Study of the Merck/INBio Agreement," *Environment* 35 (1993): 20.
26. Jeffrey A. McNeely, "Economic Incentives for Conserving Biodiversity: Lessons for Africa," *Ambio* 22 (1993): 147.
27. Malcolm W. Browne, "Folk Remedy Demand May Wipe Out Tigers," *New York Times*, September 22, 1992.
28. Personal communication, April 5, 1994. H. D. M. Moore concurs with Blackler's prediction in "*In Vitro* Fertilization and the Development of Gene Banks for Wild Mammals," *Zoological Symposium* 64 (1992): 89–99.
29. Michael E. Soulé, "Conservation Biology in the Twenty-First Century: Summary and Outlook," in *Conservation for the Twenty-First Century*, ed. David Western and Mary C. Pearl (New York: Oxford University Press, 1989), 297–303.
30. Ibid., 303.
31. Dale Jamieson, "Against Zoos," in *In Defense of Animals*, ed. Peter Singer (Oxford: Basil Blackwell, 1985), 115. What is left unstated in the disagreement between Soulé and Jamieson is a disciplinary struggle between descriptive conservation biologists and geneticists for control of the conservation agenda.
32. Thomas Eisner, "Chemical Prospecting: A Proposal for Action," in *Ecology, Economics, Ethics*, ed. Bormann and Kellert, 197.
33. Natalie Angier, "Orangutan Hybrid, Bred to Save Species, Now Seen as Pollutant," *New York Times*, February 28, 1995.
34. Ibid. Angier points out that in the United States genetic analysis has introduced similar complications into other conservation programs, including those for the red wolf, the California condor, and the Florida panther.
35. David Takacs, "Costa Rica's National Institute of Biodiversity (INBio): Biodiversidad Central," paper presented at the Nature of Science Studies Workshop, Cornell University, April 1994, 18. For "catastrophic detail," see David Takacs, *The Idea of Biodiversity: Philosophies of Paradise* (Baltimore: Johns Hopkins University Press, 1996). The new conservation initiatives waver between free-market ideology and what Andrew Ross calls "the new corporate logic of planetary management" (*Strange Weather: Culture, Science and Technology in the Age of Limits* [London: Verso, 1991], 207). Conservationists want to harness the invisible hand of the market in the service of species preservation, but at the same time they want to maintain their authority as scientists to determine which species and habits should be preserved.

36. Browne, "Folk Remedy Demand May Wipe Out Tigers."
37. Harriet Ritvo, "Race, Breed, and Myths of Origin: Chillingham Cattle as Ancient Britons," *Representations* 39 (1992): 1–22, shows the importance of prestige, purity, and "wildness" for another "charismatic" animal: beginning in the late seventeenth century, Britain's Chillingham cattle were revered for their wildness, their supposed genetic purity and distinctness from common domestic cattle (reports of miscegenation were "energetically denied" [16]), as well as their purported descendance from fashionable extinct primeval British creatures such as the "auroch" (wild ox), making them "half-legendary beasts" (6). They were extremely prestigious animals, the precious possessions only of noble families whose nature parks they inhabited.
38. McNeely, "Economic Incentives for Conserving Biodiversity," 147.
39. William Beinart, "Empire, Hunting and Ecological Change in Southern and Central Africa," *Past and Present* 128 (1990): 175–76; Donna J. Haraway, *Primate Visions: Gender, Race, and Nature in the World of Modern Science* (New York: Routledge, 1989), 26–58.
40. Takacs, "Costa Rica's National Institute of Biodiversity," 5; Ehrenfeld, "Thirty Million Cheers for Diversity." Takacs, *The Idea of Biodiversity*, treats INBio extensively.
41. Edward O. Wilson, *Biophilia* (Cambridge: Harvard University Press, 1984), 121. Similarly, David W. Orr argues, "If we complete the destruction of nature, we will have succeeded in cutting ourselves off from the source of sanity itself" ("Love It or Lose It: The Coming Biophilia Revolution," in *The Biophilia Hypothesis*, ed. Stephen R. Kellert and Edward O. Wilson [Washington, D.C.: Island Press, 1993], 437). Dick also plays with this trope, housing the humans "lucky" enough to escape the nature-depleted earth in the off-world colonies with only android slaves for companionship.

    Biophilia is a brilliant strategy for "making conservation true." Wilson is trying to make biophilia a self-fulfilling prophecy: he seeks to persuade people to behave as conservationists by convincing them that conservationism is written into their genes. To do this he necessarily dispenses with the pesky historical facts that suggest that humans have not proven to be particularly disposed toward preserving nature—quite the opposite, in fact. More perplexing still is how Wilson can believe that conservationist values are a successful evolutionary adaptation, whereas Asian values that hold that tiger parts have medicinal value are unsuccessful.
42. Roy Willis, "Cosmology, Economy, and Symbolic Loading," in *The Exploitation of Animals in Africa*, ed. Jeffrey C. Stone (Aberdeen, Tex.: Aberdeen University African Studies Group, 1988), 303–14, suggests that tribes concerned with fertility and *re*production bestow symbolic meaning on animals, whereas tribes concerned with production imbue their animals with economic value. Both value systems find expression in the discourse over endangered species in contemporary culture.
43. Dick, *Do Androids Dream of Electric Sheep?* 13.
44. Leslie Roberts, "A Genetic Survey of Vanishing Peoples," *Science* 252 (1991): 1614–17. On the Human Genome Diversity Project, see Margaret Lock, "Interrogating the Human Genome Diversity Project," *Social Science and Medicine* 39 (1994): 603–6; Elizabeth F. Drexler, "Indigenous Representations" (paper presented at the Nature of Science Studies Workshop, Cornell University, April 1994); Cori Hayden, "Our Genetic and Historic Future: Kinship, Property, and the Salvage of Endangered Genes" (paper presented at the Joint Meeting of the Society for the Social Studies of Science and the Society for the History of Technology, Charlottesville, Virginia, October 1995), and "A Biodiversity Sampler for the 1990s" (unpublished manuscript, 1995); and Jennifer E. Reardon, "From the Politics of Skin to the Politics of DNA: The Technological Transformation of Race in Post-'Race' Science" (unpublished manuscript, 1995). If the boundary between the human and the nonhuman is becoming blurred, it

is perhaps being replaced by a boundary between the civilized and the natural, a system of classification that lumps indigenous peoples in with animals and other rare "species."

45. I am adopting Peter J. Taylor's convention of using the terms "First Tier" and "Third Tier," to replace the First/Third World, Western/non-Western, and Northern/Southern antonyms, with a distinction based on class, rather than geography. I also resort frequently to the royal "we" in this essay, which I mean to refer to some generalized First Tier culture.
46. Rodrigo Gómez et al., "Costa Rica's Conservation Program and National Biodiversity Institute (InBio)," in *Biodiversity Prospecting: Using Genetic Resources for Sustainable Development*, ed. Walter V. Reid et al. (Baltimore: World Resources Institute, 1993), 63.
47. Donna J. Haraway, *Simians, Cyborgs, and Women: The Reinvention of Nature* (New York: Routledge, 1991), 230.
48. Octavia E. Butler, *Dawn: Xenogenesis* (New York: Popular Library, 1987). The quotations are from Haraway, *Simians*, 226–29 (emphasis in the original). Haraway draws an interesting parallel between Oankali gene trading and the Atlantic slave trade, but she does not connect it to the genetic surveying of indigenous people.
49. Oddly enough—or perhaps, in light of my argument, not oddly at all—these sanctions consist of excluding Taiwan, not from trade in general, but from the "legitimate" wildlife-product trade. See Thomas Friedman, "U.S. Puts Sanctions on Taiwan," *New York Times*, April 12, 1994.
50. Michael Specter, "A Too-Free Enterprise Endangers Siberian Tigers," *New York Times*, September 5, 1995.
51. John F. Burns, "Medicinal Potions May Doom Tiger to Extinction," *New York Times*, March 15, 1994.
52. See, for example, Carolyn Merchant, *The Death of Nature: Women, Ecology, and the Scientific Revolution* (San Francisco: Harper and Row, 1980).
53. Western commentators on Asia have traditionally criticized Asian cultures for not being sufficiently exploitative of natural resources. This "laziness" has led to their technological backwardness, which in turn serves as the justification for Western imperialism. See Michael Adas, *Machines as the Measure of Men: Science, Technology, and Ideologies of Western Dominance* (Ithaca, N.Y.: Cornell University Press, 1989), especially 241–58.
54. Western culture is, of course, well accustomed to portraying Asians as monsters or robots. See John Dower, *War without Mercy: Race and Power in the Pacific War* (New York: Pantheon, 1986).
55. For instance, I have a hard time believing that the tiger bone is real in the bottle of one hundred "tiger bone pills," each containing 20 percent tiger bone, that I bought (strictly for research purposes) for about five dollars in New York's Chinatown.
56. Friedman, "U.S. Puts Sanctions on Taiwan." On the discourse surrounding overpopulation, see Saul E. Halfon, "*Over*populating the World" (this volume).
57. Linden, "Tigers on Trial," 47.
58. I am indebted to Peter J. Taylor for this suggestion.
59. Dick, *Do Androids Dream of Electric Sheep?* 176.
60. See, for instance, François Delaporte, *The History of Yellow Fever: An Essay on the Birth of Tropical Medicine* (Cambridge: MIT Press, 1991), 145–46.
61. Quoted in Myrdene Anderson, "Concerning Gaia—Semiosic Production of/in/by/for Our Planet," in *Biosemiotics*, ed. Thomas A. Sebeok and Jean Umiker-Sebeok (Berlin: Mouton de Gruyter, 1992), 3.

Bubbles in the Cosmic Saucepan

Rosaleen Love

I have heard that Europe will be vanquished in the next ten thousand years. It will not be the next world war or the bombs that will ultimately destroy London, Paris, and Rome but the ice that will spread southward from the North Pole and cover the land.[1] The sea level will fall, and Atlantis may rise from the waters, a possible place of refuge for the Swiss bankers and the army generals essential for the restoration of civilization once the ice retreats.

I have also heard that the hole in the ozone layer above Tasmania is growing bigger, and that ultraviolet light will flood the earth, causing cancer in those who go out in the midday sun. Gases from the combustion of fossil fuels will rise to the upper layers of the air, raising the temperature a degree or two, melting the ice, causing the sea level to rise, flooding part of Melbourne and heralding the beginnings of the next interglacial era.

With the waters destined either to rise or to fall, migration to a mountain in Hawaii seems the answer. Summer, winter, the earth turns round the sun. Day, night, the earth spins on its axis. Or the starry sky above spins round. All is relative.

Here we are, at just one transitional stage among many, between one stage of equilibrium and the next, between a state in which the human race has flourished and the next stage in which many of the oxygen-breathing species may prove redundant to requirements, and who knows what will emerge as the new dominant life-form?

They will soon get used to it, our descendants, living on the edges of either swamps or the ice, scurrying under the feet of larger, more dominant forms of life, cane toads, or penguins, as once small ratlike creatures scurried under the clumping feet of the dinosaurs.

Twenty thousand years ago, glaciers pushed down from the mountains and the wind roared round the fortieth parallel. The seas did not totally freeze, though rippling, icy platforms spread out from the land far into the ocean, and large tabular icebergs floated by in the icy sea. The shoreline sloped under the ice to the water below, the seals chased the Adélie penguins, the whales filtered krill through their wide-open mouths, and plankton followed the flow lines in the icy oceans. Bass Strait was a green and grassy plain. People walked from Victoria to Tasmania for the summer and camped on the edge of the ice, hunting the rednecked wallabies for their skins, searching in the caves for fatty furry moths.[2]

In ten thousand years, family life will be very different, though it need not be an inevitable decline into savagery as the ice descends, the power supply fails, and the demand for scarce resources more than usually outstrips supply. The human species will probably survive, though clad in kangaroo skins sewn with needles made from the teeth of bandicoots. Camels will make a comeback, as will the wallaroos, and we shall all be hunter-gatherers again, in tune with the new environment, if we are to survive, or totally at odds with it, in which case we quite possibly will not.

There will still be food when the next Ice Age comes, and half the earth freezes, and photosynthesis occurs for the same short time in the Southern Hemisphere as it does in the north. It should take six weeks to walk from Melbourne to the rim of the ice, six weeks to get back, and six weeks to summer over where the glaciers crash down from the mountains, and the rednecked wallabies roam.

Families will gather once more around campfire middens. The children will learn how to haft stone axes with beeswax. Women will soften kangaroo tendons with their teeth, as they did in the old days. As will the men, for at least they will know more about fair shares with the division of labor between male and female and details about how teeth are formed, and the beneficial (or not so beneficial) effects of fluoride on tooth enamel.

We shall not go unprepared into the next Ice Age as they went into the last one. Everything is for the best in the best of all possible worlds, said Voltaire's Candide, until he knew better, and this takes his argument one step further into the new era. Everything is for the best in a world in which the cycles of nitrogen and carbon and hydrogen are once more operating smoothly through the realms of the animal, the mineral, and the vegetable. It may be for the best that humans are no longer in control.

It may be that the greenhouse warming of the earth will come before the next Ice Age. As the waters of Sydney Harbor rise around the waterfront at Tooronga Park, the zoo will empty its stocks to the central desert, and okapi shall graze where now the lizards dwell. Hotter, cooler, who can tell? There will come a time with cane toads born to rule, if only for the next few billion years. Why should we humans object to another highly successful life-form just because it hops along on all fours, dung-colored, with poison glands under its wart-encrusted ears? What if its sexual habits verge on the disgusting, with sometimes a preference for a road-killed partner decayed past putrefaction into desiccation? Sexual preference, too, must be subject to the changing conditions of life. Someone's putrescent cane toad is another cane toad's delight.

It may be that measurements taken over a mere one hundred years can tell something, but not enough. The last ten thousand years may show an upward trend of destruction, but what then? Change happens: take the very long-term view, and there are always variations, so that present fluctuations with the seasons are as nothing compared with the changes wrought by the coming and going of the glaciers, with their great sheets of glittering, reflective ice.

Opportunistic, entrepreneurial, the earth folds all to its capacious breast. What matters is the whole, more than the sum of its parts. The parts are interchangeable, providing they are many.

We are but bubbles in the cosmic saucepan.

The system can accommodate a lot of variation, but one day, no longer. It packs it in. It loses its amazing elasticity, quite suddenly, and slips sideways into a new way of being, overcome by forces that can no longer be accommodated. Its equilibrium has been punctuated, its status quo upset. But it still has the upper hand, though it has chosen up till now to exercise its control in ways that are relatively benign.

The universe is in the state of becoming. Quarks add to quarks, mesons to pions, neutrons to protons, and quite ordinary matter moves from the relatively simple to the absolutely complex. Nothing to begin with, then the big bang, and primitive hydrogen is formed, and more complex elements in their turn, until the interstellar gases condense and the stars shine in the sky and nucleosynthesis deep inside the outer glow forms elements more complex still, and new stars are born, collapse, and die, and in exploding carry their debris into the remoter reaches of the ever-expanding cosmos.

One day, nothing, then matter, in all its pure simplicity of form. Matter becomes more complex, and aspires to higher and higher degrees of organization. Throw ice into some water and heat it. The molecules will vibrate faster and faster, caught up in the fixed arrangements they bear to each other, but as they speed up certain fixed limits will be exceeded, and change will be fast, sudden, and discontinuous. Ice enters the new realm of water, just as a rope that has been stretched to its limit breaks to form something similar, but different. Two pieces of rope.

The amoeba has little consciousness of self. It knows only to turn to light and away from darkness. The universe has up till now had much the same reflex action to its activities. Once a higher level of organization is reached, there may well emerge a cosmic consciousness as a new factor to be taken into account in everyday life.

One day, we shall all be going about our business. The next day, there will be the voice of the aware cosmos booming in our ears, telling us what it thinks it is, and where it thinks it is going, not too worried about the sentient forms of lower life that happen to be wriggling round its interstices. The cosmic consciousness, when it comes, will probably be as concerned about us as we are about the buzzing fly or the self-important gnat.

Information overload. The voices on the radio are the background noise against which lives are lived, personal decisions are taken, interpersonal crises are daily rehearsed. The voices on the radio report what is happening, and the voices go out into the void.

What are words, but wave motion in the air, tripping out in all directions from a central point of origin? What are words but electronic messages along telecommunication channels? Electrons flow, air molecules vibrate, making a series of not so random disturbances in the information field. Billions of molecules dance to new tunes.

I hear the voices and I try to understand. Today, neatness and order. Tomorrow, chaos. That is the nature of the world. We are on the edge of the abyss, about to fall off the edge of the flat earth, to be sucked into the black hole by who knows what, some cosmic vacuum cleaner, perhaps. Who knows whose finger is on the button, ready to tidy up this earthly mess?

Better not to know.

Things when they change will change unpredictably. Look at the lesson of the dinosaurs for proof that chance and change and perhaps a sideswipe from a passing comet will tip the balance of nature into an entirely new way of being. One day everyone will be going about their business, expecting the sun to rise tomorrow. Then tomorrow, it won't happen. Everyone will be surprised and annoyed, and doubtless, like the dinosaurs, dead within days.

Deep in the jungle, the butterfly beats its wings, and the changes it creates help generate a tornado that flattens houses half a world away.

What if, though? What if, as the next stage of human evolution, a sixth sense emerges into human awareness? What if by some means people grow into a fuller sense of the electrical world around them, just as they have grown into a partial awareness of some manifestations of it, in light and sound?

What if, one day, as the world tips sideways into a new way of being, all the electricity that is leaking from the power lines and the transformers, the television towers and the satellite dishes, what if it should turn the human body electric, and change the course of evolution? Today's wild idea may be tomorrow's accepted fact. Or today's lunatic theory may be tomorrow's lunatic theory, but who will know, before tomorrow comes?

One day, some people will develop an electric sense. They will know what it is, in the way everyone knows the wind is blowing in their face, or feels the water flowing over their skin. First one person will sense it, and then another, and such is the power of a new insight, a new habit of life, such is the capacity of the human mind to adapt, to understand, that knowledge will quickly transmit itself across the globe, even into remote villages where pollution of the environment with electronic radiation has never been much of a problem. All over the world the new knowledge will soon replace the old.[3]

Of course, people will deny it, at first. Then they'll get used to it and accept it. Accept that this is the way things happen. One day, it will be forgotten that once things were different, that before the electrical machines came, things were simpler, somehow.

We shall see people in refrigerator shops, electric passion throbbing through their veins, their hair stiff with static. Shameless with it, in love with their Walkmans and their pop-up toasters. A new electric love, a love that transcends the old distinctions, drawing on a deeper kind of affinity, working at the level of the electrons, the quarks even, plugging into the electrical heartbeat at the center of creation. The private world of each individual person will link and merge into some kind of shared public electromagnetic world.

The earth will still spin around. Lines of force will sweep down from the far reaches of the cosmos, buffeted by the solar winds, dancing with the green light of the aurora, looping through the magnetosphere, traveling deep down into the earth. Along the way, waves of electric love will pass their message through various organic nervous systems, deep down into the neurons, sparking the higher levels of consciousness, uniting all in the forces of creation.

The earth will make a sideways slip, and things will be forever different. People will adapt themselves quite well to the change, and only occasionally, in the middle of the night, will someone sit up and say, "Didn't things once used to be different?"

The notion will soon be lost in the confusion of other night thoughts and will not survive into the waking day.

Notes

1. This essay originally appeared in *Arena Magazine* (Australia) 5 (June–July 1993). *Arena Magazine* is available from P.O. Box 18, North Carlton, 3054 Australia.
2. Josephine Flood, *The Moth Hunters: Aboriginal Prehistory in the Australian Alps* (Canberra: Australian Institute of Aboriginal Studies, 1980). Small brown moths,

Agrotis infusa, called Bogong by the Aboriginals, estivate in their millions in caves on the Alpine peaks of the southeastern highlands of New South Wales. In prehistoric times, the moths, fatty, furry, nutritious, and abundant, provided a valuable food supply, easily caught, easily cooked by toasting on fire, and said to be delicious. Each spring Aboriginal people traveled to the Alpine peaks from the valleys and tablelands to feast on the moths.

3. It may look as if I am referring to the human-machine interface here, and one comment has been that this final section is too laudatory of the cyborg. If my reader means that the human-machine interface is transgressed at this point in the narrative, I would say, possibly not. If humans develop a sixth sense, an electric sense, it does not necessarily mean that they may take it from the human-machine connection. They may develop an electric sense in similar ways to the electric fish of the Amazon, *Gymnotis carapo*, which use their electric sense to detect the tides, to tell the difference between their own species and others, where they are, how many, which sex, and their sexual ripeness. If anything is given a jolt in this section, it is the human/electric fish interface, and it seems to me that this can be applauded without reservation. In "A Manifesto for Cyborgs," Donna Haraway lists three crucial boundary breakdowns that the cyborg notion charts/celebrates: first, the boundary between human and animal is thoroughly breached; second, the breakdown between human/animal and machine; third, the breakdown between the physical and the nonphysical (Donna J. Haraway, *Simians, Cyborgs and Women: The Reinvention of Nature* [New York: Routledge, 1991], 151). The electricity-generating capacity of the electric fish might serve to invoke a more "natural" machine. Instead of power stations belching black smoke, there might be a more fish-friendly construction, a giant aquarium with electric fish on treadmills. Or something.

Changing Life as a set of transgressions.

Afterword: Shifting Positions for Knowing and Intervening in the Cultural Politics of the Life Sciences[1]

Peter J. Taylor

This volume, with its contributors drawn from different disciplinary persuasions within science and technology studies (STS)[2] and from geography, ecology, and developmental biology, has provided a range of interpretative angles on the metaphors, narratives, models, and practices of life sciences. *Changing Life* should help enlarge the community of participants in both cultural studies and STS and add to the emerging links between these two areas of scholarship.[3] In principle, both directions of exchange between cultural studies and STS are open to exploration. This collection, however, favors the assimilation of the study of science and technology (S&T) into cultural studies. In contrast, some practitioners within STS have argued that they have the tools needed to address culture in S&T, and have been doing so with a clarity and precision that should be more widely adopted.[4] I do not agree with this defensive STS reaction. Nevertheless, because my goal in this Afterword is to stimulate further work on problems in both STS and cultural studies, I want to provide some balance to the assimilationist tack of the volume as a whole. Let me start, therefore, from some concerns of sociology of scientific knowledge (SSK), a well-established branch of STS.

SSK shares with philosophy of science an *epistemological* concern with how scientists establish knowledge, yet it has sought to undermine philosophers' accounts by examining how scientists as actual, not idealized, agents make their science. Questions of epistemology and *agency* form my starting point here. These questions, however, become more complex than SSK tackles once we acknowledge, as *Changing Life* does, the large and heterogeneous arena in which life is (re)constructed, an arena extending from genetically hybrid organisms to transnational economies. To better address such questions, I identify five "shifts in positioning." Through the first shift, scientists come to be treated as agents who construct jointly their knowing and intervening by mobilizing heterogeneous resources; the other four shifts build on this picture. The point, however, is not to refine our accounts of how scientists work in different contexts. Instead, I want to argue that although the five shifts are already under way in particular sectors of cultural studies and STS, they can be pushed further and *applied to a wider class of agents, our-*

selves included, who not only interpret, but also intervene in science, technology, and culture.[5]

Science and Technology as Culture

Let me set a backdrop for discussing agency and positioning in both STS and cultural studies by first recapitulating the cultural-studies-assimilates-STS direction of linkage: texts and other discourse about the meaning of science and technology are a significant, even dominant part of culture. With a cultural studies orientation, one might choose some development in S&T—the more recent the better, given the field's emphasis on contemporary issues—and interpret the intertextualities in which that S&T is positioned. To a greater or lesser degree, every essay in this volume provides interpretations in this spirit: Gilbert animates our thinking with his body politic metaphors; and Gottweis encourages us to see genetic engineering as a tool for production of identity. Schroeder and Haila identify destabilizing contradictions: is ecology a resource for disciplining or for solidarity? Halfon traces discursive webs around population policy making; Taylor deconstructs global environmental discourse—and his own deconstruction. Edwards reveals a declining cultural anxiety about cyborgs; Cole points out the ironies in commercial preservation of the genetic lines. And Love deflates our overseriousness: Are we bubbles in a cosmic saucepan?

More generally, S&T is an exceptionally fertile substrate for cultural studies. The reasons are various: S&T has a history of simultaneously making universal, contextless claims and serving powerful institutions, such as the military. Cultural studies has gained much of its political purchase by demonstrating the situatedness of just such purportedly universalist or totalizing accounts and exposing the privilege such accounts afford to dominant processes and groups. "Deprivileging" requires upsetting the easy equation of Progress with Science and Technology as the producers of ever more detailed, refined accounts of nature and ever more effective interventions in nature. Cultural studies also works to develop counterdiscourses by elevating cultural strands other than the powerful and publicly visible, exploring tensions among diverse groups and identities, and focusing on conflicts during changing eras. In this spirit, dominant representations of Science and Technology can be disturbed from many angles: We can map cultural intersections across strands or sites such as those of domestic life, schooling, workplace, popular culture, and nation. We can attend to class, gender, sexual, ethnic, and other cultural differences within and among those strands. We can examine histories of traditions—ascendant, dominant, residual, subaltern, and oppositional—especially during transitions between eras, from colonial to postcolonial, or modern to postmodern. "Disturbing" can be taken further by interpreting the mutual constitution of these strands, differences, and transitions, their rhetorical and pragmatic separation, and their ongoing reconfiguration.

In short, there is readily definable work to be done in a thoroughgoing, critical cultural studies of S&T—or, rather, S&T*s*. With S&Ts assimilated into the more general category of cultures, cultural studies can provide a range of new directions of interpretation for STS.[6] Without at all intending to detract from this project, let me now consider one strand of the reverse exchange, exploring how aspects of SSK might be extended to suggest areas of future development for both STS and cultural studies.

Questions of Scientific and Cultural Agency

Sociology of scientific knowledge as it developed in the United Kingdom in the 1970s shared with philosophy of science a concern with how scientists establish knowledge claims. By looking at what scientists actually do, especially when knowledge is disputed, SSK was able to displace the idealized and agentless accounts of hypothesis testing and empirical refutation previously fashioned by philosophy of science. SSKers documented how observations and experimental demonstrations are susceptible to divergent interpretations and how this interpretative flexibility can be exploited rhetorically to maintain or dispel scientific controversy.[7] Sociology must, they argued, be brought in if the trajectories of actual and potential disputes, or, more generally, of scientific practice are to be explained.[8]

"Examine how scientists actually make their science." This directive for studying epistemology sociologically, when suitably broadened, provides a critical angle on agency in cultural studies and cultural politics. We need to clarify how agents of diverse kinds bring about the outcome under discussion, and examine how these and other agents do something with that discussion. Consider, for example, Donna Haraway's accounts of twentieth-century life sciences. During the 1980s, Haraway showed how certain episodes in various sciences involved the working out of concerns about social order and disorder. These concerns change as society changes, and this social change is, in part, conditioned by changes in the life sciences.[9] Haraway's work offered a way to extend Raymond Williams's reading of ideas of nature as reflections of ideologies of social order.[10] Not only in popular ideas about nature, but in the sciences of nature themselves, society has been naturalized, and nature socialized.[11]

Of course, reciprocal processes of naturalization and socialization occur in unevenly changing and partial ways that are sometimes contradictory.[12] Yet, where connections between science and social order are observed, the question of agency of change remains: how did scientists and allied agents actually make their work in a way that we can later interpret as corresponding to concerns about social order? Discursive interpretations of science in terms of resonances, shared metaphors, and scripts often imply that social order becomes internalized in the subjectivity of agents, and from this seat becomes expressed in all that the agents do.[13] Agents are, in this view, primarily makers and maintainers of meaning and identity. Such interpretations might suffice if, after decades have passed, a historian provides a narrative overview of the scientific and social order. However, when we attempt to extend science-social order connections up to the present and hope to intervene into future making,[14] we need more practical insight into the materiality of doing science and of doing science differently.[15]

Although SSK provided me with a starting point, I do not propose that cultural studies of S&T should model itself on SSK in order to clarify what scientific agents do in constructing scientific, technological, and cultural order.[16] After all, SSK has had little to say about the theoretical challenges that arise when the boundaries of S&T are extended well beyond the laboratory, when scientific practice includes commodification, transnational networks, regulation by states and by capital, new social movements, and discourses of globalization, marginalization, individualization, and hybridization—processes that are central to *Changing Life*.[17] Nevertheless, the epistemological bent of SSK might still be evident as I outline five moves that should help us better articulate and address, intellectually and practically, such theoretical challenges.[18]

1. From Standing on Foundations to Heterogeneous Construction

> I am studying, simultaneously, colonialist practices, engendered practices, and generational practices. I am studying unstable ecologies which are simultaneously local, regional, national, transnational, and global. I am studying the production, reproduction, consumption, and revisioning of knowledge. I am studying a community's strategic practices (visual, verbal, mathematical, mechanistic, financial, computational, institutional, pedagogic, governmental, etc.) for doing physics. I am studying how all their strategies shape and are shaped by each other. Do physicists (and historians and anthropologists) have their own kind of common sense? Are there aesthetic pleasures about thinking? Do machines crafted by physicists make science? Do rhetorical

> devices make texts? Are simulations desirable? And I am most certainly a part of what I study.[19]

Where do we position scientists as knowers and interveners? SSKers and other critical philosophers have made it impossible to imagine that scientists could stand on a firm foundation gained by accessing reality in some way independent of themselves as knowers. The fantasy of transcendent, disengaged knowledge must be replaced by recognition of necessarily partial perspectives. Those who follow the feminist-standpoint theorists prefer the partial perspectives available to women from their daily life experiences,[20] or, more generally, perspectives from subjugated standpoints, "because they seem to promise more adequate, sustained, objective, transforming accounts of the world."[21]

We can displace scientists even further from the position of neutral objectivity claimed by or for them if we redescribe their projects of representation as articulations of the "clusters of processes, subjects, objects, meanings, and commitments" that make up or situate "situated knowledges."[22] Similarly, when expressed in the terms of "actor-network" sociologists of S&T such as John Law, Michel Callon, and Bruno Latour, scientists use a wide diversity of things in the process of making S&T; they mobilize equipment, experimental protocols, citations, the support of colleagues, the reputations of laboratories, metaphors, rhetorical devices, publicity, funding, and so on.[23] The outcomes of scientists' work—theories, readings from instruments, collaborations, and so on—are accepted, in this view, because their networks of linked resources make the outcomes difficult to modify in practice. These scientific and technological outcomes become, in turn, resources for ongoing scientific work.[24]

The greater the quantity and heterogeneity of things in clusters or networks, the harder it is to pin down what knowledges *correspond to*. No one kind of thing can be separated out—and this means deep, underlying reality as well—from the list. The speed of change exacerbates the problem (I will return to this later) as does the dimension of discursivity: agents work with, among other things, images of what the world is like; faced with overlapping clusters or reticulating networks, agents discursively simplify or reduce the complexity. Moreover, they can to some extent vary the discursive reductions from context to context and employ them as additional resources in their ongoing network building. Ironing out these added twists into a single dimension of correspondence becomes at best a discursive reduction, not a plausible representational project.

Suppose we let go of questions of correspondence between knowledge and some reality, and shift our focus to the *processes* of agents building representations and other scientific products by combining a diversity of resources, that is, to the agents' *heterogeneous construction*.[25] No longer are agents persons on whom influences or factors impinge, which lead them to see, clearly or with distortion, the "nature of nature." Instead, they are persons who have to mobilize diverse resources. No longer are their subjectivities some kind of internal representation of, and thus a surrogate agent for, the external social order. They mobilize resources imaginatively, projecting themselves into possible engagements with the world in order to assess, not necessarily explicitly, the practical constraints and facilitations of establishing a scientific outcome in advance of their acting.[26] Agents' concern with modifiability of outcomes means not only that when intervening in the world they draw on representations, but also that they cannot help but implicate considerations of intervenability in their very making of representations.[27]

Clusters and networks, discursive reductions and heterogeneous constructions, imaginative representing-intervening—can this complexity be disciplined?[28] If we shift our interpretative position so as to address the processes of heterogeneous construction, epistemological challenges are raised for STS: how do we discern which of the diverse components so mobilized make a difference and analyze how those resources are combined to do so?[29] While doing this, we have to grapple with historically contingent situations resulting from multiple intersecting processes, in which boundaries and categories are problematic, levels and scales are not clearly separable, and structures are subject to restructuring. Differentiation and change, not adaptation or equilibrium, characterize these situations of "unruly" complexity.[30] The interpreter has to consider simultaneously the practical implications of different constructions and the agents' discursive reductions of those constructions. Control and generalization are difficult and no privileged standpoint exists; the boundary between scientist and engaged interpreter can hardly be maintained.

If all this were not trouble enough, as interpreters we also have to build our own networks. We select and juxtapose components in narratives, fashion boundaries and categories, and employ conventions and technologies of representation in order to convince intended audiences, to secure ongoing support from colleagues, collaborators, and institutions, and to enlist others to act on our interpretations—or, more broadly, to stimulate them to build webs that reinforce our own interpreting. In short, interpreters are also heterogeneous constructors; all the preceding discussion of how to position scientists as knowers and interveners applies to ourselves. This invites us to reflect on the range of practical conditions that enable us to build and gain support for our own interpretations of scientific activity.[31] At the same time, in most intellectual discourse it still makes practical sense for scholars to avoid such practical reflexivity and discursively discount such additional complexity.[32]

More generally, when faced with all these layers of unruly complexity, the dominant intellectual strategies have reduced or suppressed them, narrowly circumscribing some system, structure, or underlying process and representing its evolution as subject to simpler determinations.[33] Such "systemizing" is in principle rejected by cultural studies, which emphasizes contingency, contextuality, unevenness, difference, and reflexivity. In its place, cultural studies offers an image of a field of inquiry that always begins in "an in-between space where methods from existent disciplines . . . may be appropriated and refigured."[34] It "proceeds by way of a cutting-out and stitching-together of the various theories and theorists (and experiences and narratives) extracted or escaped from the various epistemological prisons. . . . This weaving together utilizes differences without isolating or 'preserving' them."[35] The propositions or "heuristics" woven together are necessarily partial; the products are contributions to ongoing weaving by others.

This alternative image of method mirrors the complexity of material and the diversity of participants in the wider arenas of sciences, technologies, and cultures with which we are concerned. What it leaves unclear, however, is just how partial propositions can be woven together while interpreters remain in the "in-between space." As a consequence, simple determinations and correspondences remain an implicit resource in the metaphor-making and oppositional strategies of cultural studies. Let me amplify this criticism by describing further shifts that I associate with the shift to heterogeneous constructionism.

2. *From Mental and Verbal Images to "Acting as if" as Metametaphor*

Among the resources drawn on by agents, whether they are working as scientists or cultural interpreters, is a sense of what the material and social world is like. Interpreters observing this "likening" try to convey what *it* is like, that is, they employ "metametaphors." Consider, for example, the range of terms that involve likening in Haraway's work: agents are involved in noninnocent conversations using visualization technologies; holding partial perspectives; situated knowing; storytelling; embodied vision; worldly diffraction; viewing the doings; materialized refiguration. Without worrying what each of these terms means, notice that all but the last term connote the making of mental and verbal images, images that we *believe* or *think* the world is like, or that we *speak* or *write* as if it were like. In general, whether one highlights the public and interactive dimension of representation, as Haraway (and cultural studies) does, or its mentalistic and subjective connotations, likening is taken to mean having an image that corresponds to the world.

Yet there is another sense of likening, one that parallels the first shift in our basis for knowing and intervening, namely, we also *act as if*. In fact, although viewing, speaking, writing, and thinking are indeed actions, they are particular kinds; "acting as if" could be viewed as a more inclusive metametaphor of likening.[36] Proceeding in this way, when we examine what is at stake in any representational or discursive work, we would ask what it is that the agents are trying with their ideas to *do* something about and what needs to be done *practically* in order to modify their moves. Following this line of questioning, the activity of scientific and interpretative agents can be interpreted as richly metaphorical in the more inclusive, "acting as if," sense.[37] The global computer modelers in my essay on global environmental discourse, for example, think that human activity forms a system to be managed. More important, however, their categories, tools, diagrammatic conventions, gaming, and social positioning jointly enable them to act—actually or in powerful fantasies—as if they were the planet's managers (or, at least, their close advisers).[38]

3. *From an "Existence" Imaginary to "Construction" Work*

In mathematics, a distinction is made between existence and construction theorems. The former do their work by demonstrating, for example, that a system of equations governing some process will have one and only one point of equilibrium. A construction theorem takes on the more difficult task of laying out the steps or procedure needed to find that point. By analogy, early SSK tended to work within an existence imaginary. The existence of interpretative flexibility, for example, counters philosophical claims that, by rational procedures alone, scientists can allow nature to adjudicate among scientific claims. SSK tells us little, however, about how those particular claims came to exist rather than others. Not surprisingly, although SSK disturbs philosophers and scientists who had wanted to retain authority over representing scientific method, it rarely stimulates scientists to do science differently.

Similarly, cultural studies of S&T has thus far carried out its important oppositional work more within an existence imaginary than through "construction work." Against universalist accounts, cultural analysts argue that knowledge *is* situated, that there *is* a multiplicity of necessarily partial perspectives and voices implicated in S&T. Against simple determinations and firm foundations in reality, we are shown that there *is* a "mess"[39] of scientific practices and discourses taking place at sites that range from the psyche to the international political economy.[40] Ironic inversions upset literal interpretations and simple moral lessons; hybrids transgress foundational categories; science that promises enlightenment and liberation spawns confusions.[41]

In the different ways it points to the messy complexity of S&T's situatedness, cultural studies of S&T has already succeeded in disturbing people from both the SSK and scientific communities. From the defensive and even angry responses to cultural studies of S&T, it is clear that the SSKers and scientists alike had wanted to retain authority over representing scientific practice.[42] Yet, disturbing people is one thing; by what steps and practices, however, are science and the culture of science to be changed? The move from cultural studies of science to cultural *politics* is not straightforward. Who are the intended agents of change in a cultural politics of science? What are critical interpretative agents supposed to do with and through situating, cultural interpretation? If scientists are to be drawn into critical interpretative collaborations, how is this to be achieved? The explicit and implicit answers so far to these questions tend not to do justice to the situatedness (or heterogeneous constructedness) being interpreted. Nor do they make much of the warrant for practical reflexivity that follows from recognition of the situatedness of any interpretation.[43] Let me develop some examples outside this collection to tease out this assessment.

Andrew Ross invites us to join "a reasoned public discussion of issues concerning S&T," through which we expose the connections that exist among social, natural, and economic life, and articulate the different connections we might desire.[44] Although public discussion of S&T is already and increasingly multivocal, Ross reminds us of the deep logic behind "technoculture": "capitalist reason, not technical reason, is still the order of the day."[45] By implication, socialists and others who see themselves resisting capitalist logic should be especially qualified and motivated to reason and discuss. David Hess, an anthropologist of S&T, also draws attention to techno(multi)culturalism. To challenge the power of the dominant social order and the S&T that contributes to it, he promotes solidarity with socially and scientifically marginalized groups. In particular, he argues, we should give attention to heterodox science and knowledge systems. Teaching about the multiculturalism of S&T should, moreover, help recruit and retain people from groups underrepresented in technical professions, and this would lead to different S&T.[46]

I have extracted these simple themes about agency and cultural politics from introductory and concluding material in books by Ross and Hess. Although the body of their texts presents more complex pictures, their introductions or conclusions can sensitize us to the emergence of simple themes within more complex pictures and to the use of such themes as interpretative resources. In this spirit, let us note some discursive reductions and hidden determinisms in the accounts of some of the most complexity-embracing of writers, Traweek and Haraway.

Traweek and Haraway amplify Ross's and Hess's ideal of a wider, multivocal discussion about technoculture. At the same time, they complicate oppositional solidarities by highlighting transgressions across the boundaries between marginal and dominant formations. Traweek, for example, describes a Japanese woman scientist in the male- and U.S.-dominated field of high energy physics who used her experience in big multinational collaborations to identify "gleanings" of data left after the big boys took what they wanted. This woman then arranged a mutually beneficial deal with a computer company so that she could build the equipment she needed to analyze those data.[47] However, within such stories of ambiguity lie some simpler themes.

Traweek associates her mess with a "phase transition" between eras.[48] Knowledges, technologies, and societies have been based on simplicity, stabilities, uniformities, taxonomies, regularities, and hierarchies. Now we are facing complexity, instabilities, variations, transformations, irregularities, and diversity. Sometimes, however, Traweek presents the second set of attributes as the way the world has always been; what has changed is the favored aesthetic of representation. I think we can attribute this equivocation to the greater rhetorical power of the claim that a marked (evolutionary) transition is occurring (to complexity, postmodernism, etc.). To grab our attention, to stimulate us to respond, it seems enough for Traweek to point to the new era of complexity (or to point to transgressions as evidence of its coming into existence). In contrast, if we followed her shift in aesthetics interpretation, we would need to analyze the ongoing reconfiguration of knowledges, technologies, societies, and aesthetics in order to identify where and how to intervene. The mere existence of a transition provides little insight into how to pursue this more difficult construction work.

The cultural politics to which Haraway's transgressors lead us is also interestingly ambiguous. Harvard patents a mouse that is transgenic, corporate and academic, natural and commodity, organic and technical—a cyborg.[49] Some cyborgs warrant scrutiny, especially those originally designed for war making. Other cyborgs provide a transformative standpoint; like their kin, the Sister Outsider, Inappropriate(d) Other, Coyote Trickster, and FemaleMan©, they are the marginal, dominated, silenced outsiders who move into areas previously off-limits.[50] However, if Sister Outsiders provide special standpoints, why privilege situations in which they playfully, transgressingly negotiate change? Is transgression *good*? Why rule out havens or places of refuge, outside the dominant mess, in nurturing, organic communities? After all, Haraway's favoring of Sister Outsiders has helped enlist the allegiance of many who desire some solidarity and self-esteem in their marginal positions.

We cannot find refuge in an organic unity for reasons that depend—at least, in my reading of Haraway—on *inexorable commodification*.[51] Once a market is created where there was not one before, more and more people's lives are transformed by production and exchange of the commodity; there is no going back. Because we cannot escape this, Haraway invites us to become more self-consciously implicated in the process. But how are we to join in the market in ways that allow us to distinguish resistance from accommodation? Although this question is not clearly answered in Haraway's complex accounts of "material-semiotic" production,[52] the implicit theme of inexorable commodification recalls a Marxist economic determinism, which would direct resisters to the necessity of class struggle. Yet the call to class struggle assumes that many differences among agents can be subordinated in the cause of more effective struggle against or resistance to dominant economic structures.

Haraway, Ross, Hess, and Traweek would not deny the limitations of class- or solidarity-based politics and are quite sensitive to the multiple dimensions of difference. Nevertheless, accounts that point to the existence of differences, ironies, transgressions, and other aspects of unruly complexity can still build on or build in complexity-collapsing concepts of politics. Whether this is one resource among many or dominates the accounts, there remains in cultural politics room for much more construction work.

4. *To Intermediate Complexity between Systems or Unruly Complexity*

Traweek's mess, Haraway's clusters, and my unruly complexity are like supersaturated solutions; any object placed in them initiates precipitation. We could, in an effort to preempt discursive reductions and subvert any recourse to an unacknowledged determinism, try to lower into the solution some "grid" of *intermediate complexity*. We would want material to crystallize out of solution simultaneously along a distinct set of directions and along the interlinkages among those strands.[53] This combination would allow us to trace out the implications of the *intersections* of these strands. Let me make intermediate complexity concrete by continuing on the theme of commodification.

Commodification does appear almost inexorable; reversals seem rare. But suppose we take this putative determinism as something in need of explanation. After all, in a world of heterogeneous construction, we would expect commodification to be orchestrated, contested, and, at times, thwarted. This is evident in Richard A. Schroeder's account (in this volume) of competing efforts to develop market gardens and fruit tree orchards ("Contradictions along the Commodity Road to Environmental Stabilization"). Similarly, in another *political-ecological* analysis, undertaken by a Mexican colleague, Raúl García-Barrios, and his brother Luis, rich caciques supervised a stable moral economy of norms and reciprocal expectations among unequal, cooperating agents, and a stable agro-ecology of hillside terracing during the nineteenth century in villages in Oaxaca, Mexico. This system both depended on and made possible their keeping the villages isolated from markets.[54]

The challenge that such political-ecological analyses allow us to articulate better is exposing the *construction* of commodities, as well as the resistance to and reversals of such construction. The breakdown of this Oaxacan agro-ecology after the Mexican revolution is instructive. The revolution ruptured the moral economy by taking away the power of the caciques. Following peasant migration to industrial areas and semiproletarianization of the rural population, village transactions became monetarized, and the collective institutions collapsed. The terraces began to erode. Goat herding, which was taken up because of its low labor requirements and was not regulated by strong local institutions, exacerbated the erosion. National food-pricing policies that favored urban consumers meant that corn was grown in the villages only for subsistence needs; maize remains to this day uncommodified.

Even in the very condensed sketch in figure 1, we can discern the intersection of processes operating at different spatial and temporal scales, involving elements as diverse as the local climate and geomorphology, social norms, work relations, and national political economic policy. No one kind of thing, no single strand on its own could be sufficient to explain the curently eroded hillsides. This contrasts with competing explanations that center on a single "dynamic"—for example, population growth—as the cause of environmental change, or the power of capitalism and commodification to penetrate local economies. In this sketch, I have also stepped away from debates centered around big oppositions, such as ecological versus economic rationality, or critical realism versus social constructivism.

While avoiding underlying determinisms and big oppositions, an account that identifies definite processes, which are then presented as intersecting, does abstract away some of the unruly complexity. The result is an intermediate complexity, which has implications for how one responds to potential commodification and, in the Oaxacan case, to environmental degrada-

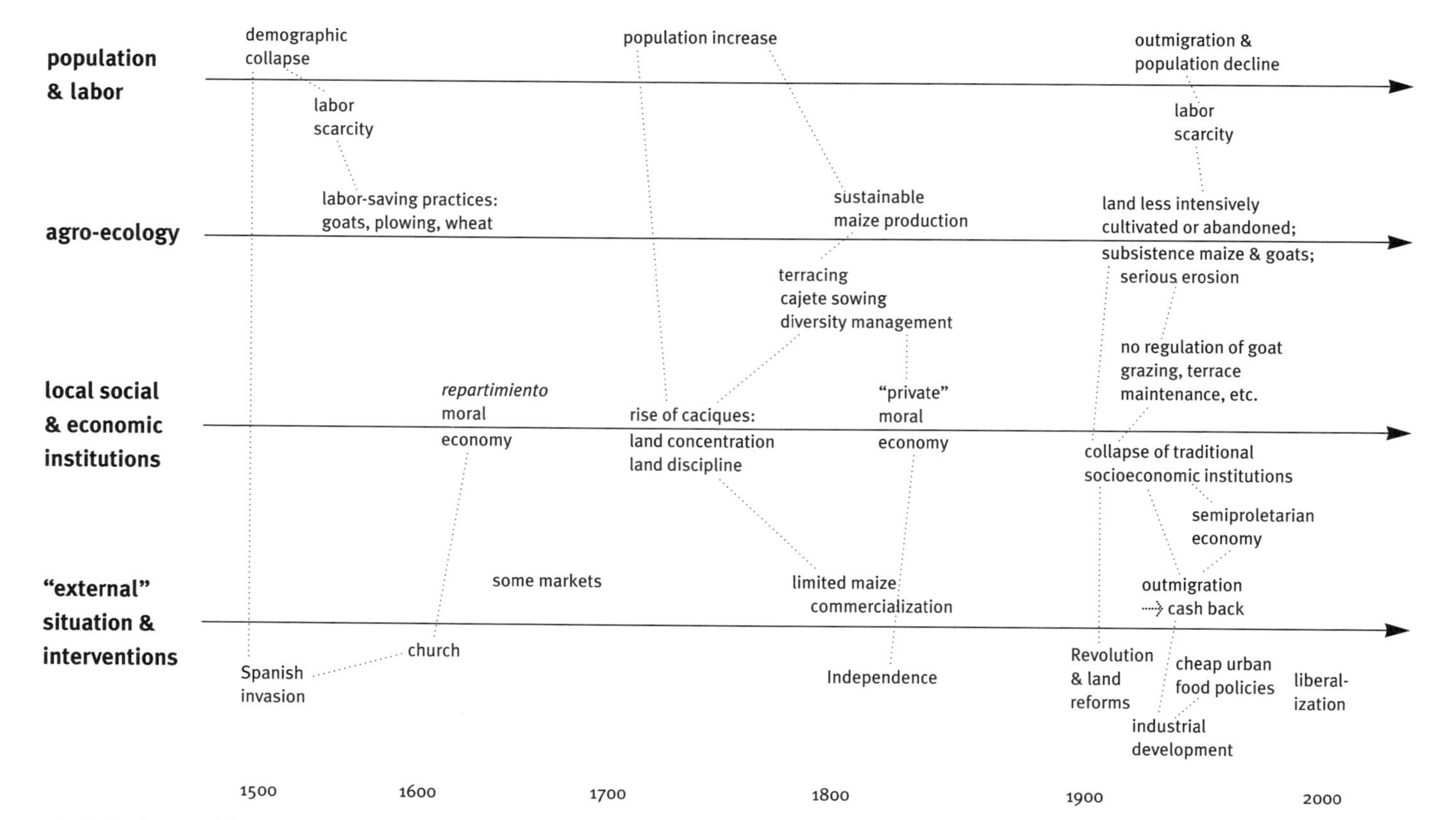

Fig. 1. A schema of the socionatural intersecting processes leading to soil erosion in San Andrés, Oaxaca. The dotted lines indicate connections across the different strands.

tion. The intersecting processes account does not, for example, support government or social movement policies based on simple themes, such as economic modernization by market liberalization, sustainable development through promotion of traditional agricultural practices, or mass mobilization to overthrow capitalism. Instead, it highlights the opportunities for *linking* multiple, smaller, and thus doable interventions within the intersecting processes. However, to gain support for these interventions, including support in the form of linkage to other interventions, one would have to cultivate particular institutional and personal resources, agendas, and alliances. By taking these particularities into account, it might be objected that practically reflexive intermediate complexity would always dissolve back into some supersaturated solution. Possibly, but this remains to be seen—or, rather, to be worked out.[55]

5. *From Increasing to Differential Speed and Extent*

Schroeder's account of gardens and trees in The Gambia is also an account of agents who negotiate processes and mobilize resources that span the local and transnational, material and discursive, traditional and innovative. This picture will seem familiar to those in cultural studies who relate their subjects to the changing global or transnational order. Discussions of changes in the "new world dis/order" highlight the increasing extent of economic and cultural connections or, complementarily, their increasing speed. In technoculture, the icons of extent and speed are the Internet and the ever-accelerating project of genome sequencing.[56] In this context, I want to suggest one last shift of positioning.

Consider William Cronon's widely read account of the nineteenth-century emergence of a "Metropolis of Nature," the city of Chicago.[57] The picture he presents is of ever-increasing speed and expanding extent. However, the motor of the changing capitalism he describes is not simply speed and increasing extent, but *differential* speed and extent. The futures market, for example, takes off not simply because telegraphic communication connects the world more rapidly, but because some people in Chicago have access to that information well before and in greater detail than, say, farmers. Differentials provide a purchase for commodification processes in general. At the same time, they provide us a purchase for exposing the different and differentiating agents implicated in the orchestration and contestation of those processes.

By attending to differentials and differentiation, cultural studies can avoid static notions of difference or underlying determinisms. Moreover, this emphasis is completely consistent with the idea that boundaries will be problematic, transgressions will be abundant, and processes of different scales will intersect. It is also consistent with the following, quite difficult challenge ahead for the cultural politics of S&T: to position different scientists and to position our different selves as knowers and interveners, we need ways to work with the unequal and heterogeneous practical and conceptual facilitations of sciences, technologies, and cultures.

A Rough, Ongoing Construction

The shifts involved in moving onto a terrain in which we address heterogeneous construction, acting as if, construction work, intermediate complexity, and differentials are, as I mentioned earlier, already under way in both cultural studies and STS. This Afterword promotes these shifts, yet remains firmly within an existence imaginary. It does not clarify the intended agents of change in any specific case, nor the steps or practices through which the science, technology, and culture are to be changed.

Taking into account the shifts of positioning still under way, the volume as a whole emerges as necessarily a rough construction. The Afterword resists the understandable tendency editors have to smooth over such roughness. Instead, I hope that readers position the work here as contributions, in need of clarification and extension, to intersecting projects in development. As declared in the Introduction, the contributors to *Changing Life* seek, in a spirit of necessary partiality, to join with others changing life in a changing social (dis)order. We hope that, through the many and diverse resources this collection provides, we are contributing to diverse interventions into processes linking genomes, ecologies, bodies, and commodities.[58]

Notes

1. I gratefully acknowledge the comments and suggestions of Ann Blum, Paul Edwards, Saul Halfon, Stefan Helmreich, Bill Lynch, Andrew Ross, and Joe Rouse in response to drafts of this essay.
2. When I use the label STS and refer to its constituent disciplines, I am thinking mostly of the academic fields of sociology, history, anthropology, politics, and philosophy of science and technology. One of the aims of this essay and the volume as a whole is, however, to expand the range of scholars who identify with the STS label. And similarly for the label of cultural studies (see the next section titled "Science and Technology as Culture" for an overview of what I consider to be the salient characteristics of cultural studies). I intend my points to be relevant and challenging generally to scholars who identify with one or both of these areas. At the same time, I realize that

specific sectors of STS and cultural studies have in their own ways already tackled some of the issues in greater depth than I can do justice to here.

3. Other recent anthologies linking cultural studies and STS include Stanley Aronowitz, Barbara Martinsons, and Michael Menser, eds., *Technoscience and Cyberculture* (New York: Routledge, 1996); Chris H. Gray, ed., *The Cyborg Handbook* (New York: Routledge, 1995); Gary Downey, Joseph Dumit, and Sharon Traweek, eds., *Cyborgs and Citadels: Anthropological Interventions on the Borderlands of Technoscience* (Seattle: University of Washington Press, 1997). We acknowledge in the Introduction to this volume the leadership role of Donna Haraway and Sharon Traweek in developing the exchange between STS and cultural studies. I take the liberty in this Afterword of referring, sometimes critically, to their work to help me articulate the shifts that have been made and the further shifts I propose.
4. Harry M. Collins, "Review of Bruno Latour, *We Have Never Been Modern*," *Isis* 85, no. 4 (1994): 672–74; Peter R. Dear, "Cultural History of Science: An Overview with Reflections," *Science, Technology & Human Values* 20, no. 2 (1995): 150–70.
5. In this light, I use the term "epistemology" to refer to concerns about how various agents, not only scientists, establish knowledge claims and interpretations. Moreover, as will become evident, knowledge making is treated as inextricably connected with a wide range of other practices.
6. For alternative descriptions of cultural studies as this field relates to science, see Sharon Traweek, "Introduction to the Cultural and Social Studies of Sciences and Technologies," *Culture, Medicine, and Psychiatry* 17 (1993): 3–25; David Hess, *Science and Technology in a Multicultural World: The Cultural Politics of Facts and Artifacts* (New York: Columbia University Press, 1995); Michael Menser and Stanley Aronowitz, "On Cultural Studies, Science, and Technology," in *Technoscience and Cyberculture*, ed. Aronowitz et al., 7–28; Gary Downey, Joseph Dumit, and Sharon Traweek, "Locating and Intervening," in *Cyborgs and Citadels*, ed. Downey et al.; Joseph Rouse, "What Are Cultural Studies of Scientific Knowledge?" *Configurations* 1, no. 1 (1992–93): 1–22, and *Engaging Science: How to Understand Its Practices Philosophically* (Ithaca, N.Y.: Cornell University Press, 1996).
7. Harry M. Collins, "Stages in the Empirical Programme of Relativism," *Social Studies of Science* 11 (1981): 3–10.
8. Some scholars in STS and cultural studies oppose the goal of explanation. See Rouse, "What Are Cultural Studies of Scientific Knowledge?", and *Engaging Science*; Hess, *Science and Technology in a Multicultural World*; and Bruno Latour, "The Politics of Explanation: An Alternative," in *Knowledge and Reflexivity: New Frontiers in the Sociology of Knowledge*, ed. Steve Woolgar (London: Sage, 1988), 155–76. What they oppose is only one particular form of explanation, in which a range of outcomes in one variable realm (e.g., science) is tied to some feature of a relatively stable realm (society). This opposition is based on *(a)* (correctly in my view) not seeing social life as stable or as a realm separate from science, and *(b)* wanting to highlight the novel coalitions and outcomes involved in the production of science and society. However, the accounts of networks of resources or webs of meaning that they advocate build on multiple, diverse causes; see, for example, Bruno Latour, *Science in Action: How to Follow Scientists and Engineers through Society* (Milton Keynes: Open University Press, 1987). Rouse and Hess also want to shift explanation out of the center of STS's focus (and perhaps out of the picture altogether), arguing that this is necessary in order to make room for reflexive, politically engaged practice. I agree with this last goal, but prefer to rework explanatory practice rather than to act as if one were abandoning it. For further discussion of these points, see later in this essay and the appendices to Peter J. Taylor, "Building on Construction: An Exploration of Hetero-

geneous Constructionism, Using an Analogy from Psychology and a Sketch from Socio-Economic Modeling," *Perspectives on Science* 3, no. 1 (1995): 66–98.

9. Donna J. Haraway, "High Cost of Information in Post-World War II Evolutionary Biology: Ergonomics, Semiotics, and the Sociobiology of Communication Systems," *Philosophical Forum* 13, nos. 2–3 (1981–82): 244–79; Donna J. Haraway, "Signs of Dominance: From a Physiology to a Cybernetics of Primate Society," *Studies in History of Biology* 6 (1983): 129–219; Donna J. Haraway, "Teddy Bear Patriarchy: Taxidermy in the Garden of Eden, New York City, 1908–1936," *Social Text* 11 (1984–85): 20–64.
10. Raymond Williams, "Ideas of Nature," in *Problems in Materialism and Culture* (London: Verso, 1980), 67–85. It should be noted that Williams was, in most accounts, one of the important early figures in British cultural studies.
11. Of the essays in this volume, see in particular those of Gilbert, "Bodies of Knowledge," and Haila, "Discipline or Solidarity?"
12. For example, organismic metaphors in social and biological thought gave way after World War II to both cybernetic and individualistic metaphors. See Haraway, "High Cost of Information in Post-World War II Evolutionary Biology"; Peter J. Taylor, "Technocratic Optimism, H. T. Odum, and the Partial Transformation of Ecological Metaphor after World War II," *Journal of the History of Biology* 21, no. 2 (1988): 213–44; Gregg Mitman, "Defining the Organism in the Welfare State: The Politics of Individuality in American Culture, 1890–1950," *Social Sciences Yearbook* 18 (1994): 249–80. In the life sciences, ecology has come to coexist with both the environmental movement and the industry of environmental management. In social thought more generally, the organic community under threat during the Great Depression was eclipsed by post-World War II optimism about preventing "violent oscillations" through efficient systems of feedback, communication, and command/control. This optimism is now tempered, yet life continues to be reconstructed, ever more intimately and extensively.
13. See, for example, Stefan Helmreich, "Replicating Reproduction in Artificial Life; or, the Essence of Life in the Age of Virtual Electronic Reproduction," in *Reproducing Reproduction*, ed. Sarah Franklin and Helena Ragoné (Philadelphia: University of Pennsylvania Press, 1997); Bill Lynch and Joe Rouse (personal communication; see also Rouse, *Engaging Science*) reminded me that an alternative reading of these discursive interpretations is that shared metaphors and so on are just those aspects of language, conceived of as an extraindividual phenomenon, that people know how to react to. In principle, people do not have to be seen as carrying the metaphors around inside their heads. A combination of or equivocation between these two readings is, for example, evident in Lily Kay, "A Book of Life? How a Genetic Code Became a Language," in *Controlling Our Destinies*, ed. Philip Sloan (Notre Dame: University of Notre Dame Press, forthcoming). In practice, I believe, many interpreters of science fit my reading better than Lynch and Rouse's. This can be the case, moreover, even for interpreters who insist on the emergence of meaning from the patterns and messiness of interactions among agents. When they omit discussion of how agents actually make their work, their accounts also become readable in terms of society-internalized-in-subjectivity. See the discussion of metametaphors later in this essay.
14. This was one of the motivating themes of the conference sessions from which this volume originated on the topic "Changing Life in the New World Dis/order." These sessions, co-organized by Irving Elichirigoity and myself, were held at the July 1993 meetings of the International Society for History, Philosophy, and Social Studies of Biology, Brandeis University.
15. From the cultural studies angle, there might seem to be little reason to keep the focus on scientists, or, at least, not on mainstream scientists. Abundant material is provided by examining S&T as they are invoked in wider cultural arenas by diverse social groups.

See Andrew Ross, *Strange Weather: Culture, Science, and Technology in the Age of Limits* (London: Verso, 1991); Hess, *Science and Technology in a Multicultural World*. I note, however, that when cultural analysts of S&T attend to the reception of science more than to its production, they risk implying that scientists' practice and theories are indefinitely malleable. In any case, in what follows I will shortly extend my category of agents to include interpreters as well as scientists.

16. Dear, "Cultural History of Science"; Steven Shapin, *Social History of Truth: Civility and Science in Seventeenth-Century England* (Chicago: University of Chicago Press, 1994).
17. SSK has been most illuminating when focused on specific sites and junctures, especially laboratory experiments and other disputes over the reliability of knowledge. Regulation, new social movements, and transnational discourses do enter the SSK work of, for example, Sheila Jasanoff, Brian Wynne, and Steven Yearley; see Sheila S. Jasanoff, Gerald E. Markle, James C. Petersen, and Trevor J. Pinch, eds., *Handbook of Science and Technology Studies* (Thousand Oaks, Calif.: Sage, 1995). The historical work of Simon Schaffer and his students at Cambridge University might—depending on one's definition of SSK—also be subverting my generalization about SSK; see, for example, Simon Schaffer, "Babbage's Intelligence: Calculating Engines and the Factory System," *Critical Inquiry* 21, no. 1 (1994): 203–27.
18. Epistemology is construed here in the broad sense described in note 5.
19. Sharon Traweek, "Worldly Diffractions: Feminist and Cultural Studies of Science, Technology, and Medicine," presented at the annual meeting of the American Sociological Association, Los Angeles, August 1994. Traweek is describing the study of high-energy physics communities in Japan and the United States.
20. Sandra Harding, *Whose Science? Whose Knowledge? Thinking from Women's Lives* (Ithaca, N.Y.: Cornell University Press, 1991); Sergio Sismondo, *Science without Myth: On Constructions, Reality, and Social Knowledge* (Albany: State University of New York Press, 1996).
21. Donna J. Haraway, "Situated Knowledges: The Science Question in Feminism and the Privilege of Partial Perspective," *Feminist Studies* 14, no. 3 (1988): 584. Sociologist of science Thomas Gieryn has observed that Auguste Comte developed a working-class standpoint theory ("Objectivity for These Times," *Perspectives on Science* 2 [1994]: 324–49). Comte proposed, for example, that "The working class is better qualified than any other for understanding, and still more for sympathizing with, the highest truths of morality" (Gertrud Lenzer, ed., *Auguste Comte and Positivism: The Essential Writings* [New York: Harper and Row, 1975], 351).
22. Donna J. Haraway, "Mice into Wormholes: A Technoscience Fugue in Two Parts," in *Cyborgs and Citadels*, ed. Downey et al. See also Haraway, "Situated Knowledges," 575–99.
23. Latour, *Science in Action*; John Law, "Technology and Heterogeneous Engineering: The Case of Portuguese Expansion," in *The Social Construction of Technological Systems: New Directions in the Sociology and History of Technology*, ed. Wiebe E. Bijker, Thomas P. Hughes, and Trevor J. Pinch (Cambridge: MIT Press, 1987), 111–34.
24. Unlike Callon, Latour, and others, I will not embrace nonhumans in our discussion of agency, because this move tends to reduce human agency to a lowest common denominator. See Peter J. Taylor, "What's (Not) in the Mind of Scientific Agents? Implicit Psychological Models and Social Theory in Social Studies of Science," paper presented at the annual meeting of the Society for Social Studies of Science, West Lafayette, Indiana, November 1993. The resistance of nonliving things and the agency of nonhuman organisms can be addressed in terms of the difficulty human agents have in mobilizing resources.
25. My use of this loaded term is intended to preserve connotations of construction as a

process of building from materials, but to downplay connotations of constructions as ideas reflecting or corresponding to some social position; see Taylor, "Building on Construction." I am not, as it is still obligatory to note, advocating unbridled relativism. I accept that close correspondence between knowledge and some underlying reality can be a significant resource in heterogeneous construction. The point, however, is that practice is never determined by any single kind of resource.

26. Associating imagination and the labor process is an idea of Marx. See *Capital*, vol. 1, Part III, chapter 7, section 1, reprinted, for example, in Robert C. Tucker, ed., *The Marx-Engels Reader* (New York: Norton, 1978), 344–45. For a relevant discussion of this passage, see Stephen Robinson, "The Art of the Possible," *Radical Science Journal* 15 (1984): 122–48.
27. Peter J. Taylor, "Re/constructing Socio-Ecologies: System Dynamics Modeling of Nomadic Pastoralists in Sub-Saharan Africa," in *The Right Tools for the Job: At Work in Twentieth-Century Life Sciences*, ed. Adele Clarke and Joan Fujimura (Princeton, N.J.: Princeton University Press, 1992), 115–48; Taylor, "Building on Construction."
28. We should not be using "complexity" without discussing its recent history (see Hess, *Science and Technology in a Multicultural World*, 106–16, for a schematic overview), which centers around nonlinear dynamics and cellular automata; George Cowan, David Pines, and David Meltzer, eds., *Complexity: Metaphors, Models, and Reality* (Reading, Mass.: Addison-Wesley, 1994). I will, however, only note that the picture of "unruly complexity" to follow fits neither of two alternative foundational principles for theories of complexity: simple rules lead to complex behaviors, or macroregularities can arise statistically from large numbers of similar entities.
29. Taylor, "Re/constructing Socio-Ecologies" and "Building on Construction." Other scholars take this complexity as a warrant for breaking away from epistemological concerns, or, at least, from the associations attached to the label epistemology; see Rouse, *Engaging Science*.
30. This picture is developed in the context of social-ecological relations in Peter J. Taylor and Raúl García-Barrios, "The Social Analysis of Ecological Change: From Systems to Intersecting Processes," *Social Science Information* 34, no. 1 (1995): 5–30.
31. See my essay "How Do We Know We Have Global Environmental Problems?" in this volume. The call for reflexivity regarding STS's own interpretations has been a theme in sociology of science for more than a decade; see Steve Woolgar, ed., *Knowledge and Reflexivity: New Frontiers in the Sociology of Knowledge* (London: Sage, 1988); Malcom Ashmore, *The Reflexive Thesis: Wrighting Sociology of Scientific Knowledge* (Chicago: University of Chicago Press, 1989). The emphasis in reflexive STS (and cultural studies), however, has been on the textual and rhetorical strategies used to advance an argument or interpretation. Downey et al., "Introduction" to *Cyborgs and Citadels*, and Menser and Aronowitz, "On Cultural Studies, Science, and Technology," also promote reflexivity about unraveling the "knots" of technoculture.
32. See, e.g., Shapin, *Social History of Truth*, xv.
33. The anthropologist Eric Wolf critiques this in his *Europe and the People without History* (Berkeley: University of California Press, 1982), 385–91: "Societies emerge as changing alignments of social groups, segments, and classes, without either fixed boundaries or stable internal constitutions. . . . Therefore, instead of assuming transgenerational continuity, institutional stability, and normative consensus, we must treat these as problematic. We need to understand such characteristics historically, to note the conditions for their emergence, maintenance and abrogation" (387).
34. Menser and Aronowitz, "On Cultural Studies, Science, and Technology," 17.
35. Ibid., 24.
36. Timothy Mitchell identifies a master metaphor in social theory, the distinction be-

tween persuading and coercing. He observes that this dualism, which opposes meaning to material reality, underwrites most strategies of power; Timothy Mitchell, "Everyday Metaphors of Power," *Theory and Society* 19 (1990): 545–77. In STS and cultural studies, an analogous deep split is that between believing and acting, representing and intervening. Mitchell's account of this master metaphor's persistence might be read as a comment on the difficulties of shifting from mental and verbal images to "acting as if" as an interpretative metametaphor: "One [reason for the persistence] stems from the fact that [the master metaphor] is indissociable from our everyday conception of person[s] . . . as unique self-constituting consciousnesses living inside physically manufactured bodies. As something self-formed, this consciousness is the site of an original autonomy . . . [that] defies the way we think of coercion. It obliges us to imagine the exercise of power as an external process that can coerce the behavior of the body without necessarily penetrating and controlling the mind" (545).

37. To a large degree this point matches the emphasis of Haraway and others in cultural studies of S&T on the relationship of metaphors to concrete ways of dealing with things (thus her "material-semiotic actors"; see Haraway, "Situated Knowledges"). The choice of metaphor for likening, however, suggests that more work is needed to move beyond mental and verbal construals of metaphor and of action.

    In my reading, the literature that analyzes in general (as against in specific instances) the use of metaphors has been dominated by three related metametaphors: (1) metaphors are root, fundamental, underlying things that shape the surface layers; (2) mental things—thoughts, expectations, what we see—shape our actions; and (3) culture or society gets into these thoughts (and so we can be taught how to conceive/perceive the world). These metametaphors are not helpful for developing the idea that all action and thought is constructed in practical activity from heterogeneous resources. For examples of the dominant metametaphors, see Kurt Danzinger, "Generative Metaphor and the History of Psychological Discourse," and Kenneth J. Gergen, "Metaphor, Metatheory, and the Social World," in *Metaphors in the History of Psychology*, ed. David E. Leary (Cambridge: Cambridge University Press, 1990), 331–56 and 267–99, respectively; George Lakoff, "The Contemporary Theory of Metaphor," and Michael Reddy, "The Conduit Metaphor: A Case of Frame Conflict in Our Language about Language," in *Metaphor and Thought*, ed. Andrew Ortony (Cambridge: Cambridge University Press, 1993), 202–51 and 164–201, respectively; Nancy Leys Stepan, "Race and Gender: The Role of Analogy in Science," *Isis* 77 (1986): 261–77.

38. Taylor, "Technocratic Optimism" and "Re/constructing Socio-Ecologies"; Peter J. Taylor and Ann S. Blum, "Ecosystems as Circuits: Diagrams and the Limits of Physical Analogies," *Biology & Philosophy* 6 (1991): 275–94.

39. Traweek, "Worldly Diffractions."

40. See the quote from Traweek at the beginning of the section titled "From Standing on Foundations to Heterogeneous Construction."

41. See, for example, the essays in this volume by Simon Cole, "Do Androids Pulverize Tiger Bones to Use as Aphrodisiacs?" and Rosaleen Love, "Bubbles in the Cosmic Saucepan." In fact, the force of the existence imaginary is evident in most of the essays of *Changing Life*.

42. On the SSK side, see Collins, "Review of Bruno Latour"; Dear, "Cultural History of Science"; and Trevor J. Pinch, "Review of Hess and Layne, *The Anthropology of Science and Technology*," *Isis* 86, no. 2 (1995): 358. On the scientist side, see Paul R. Gross and Norman Levitt, *Higher Superstition: The Academic Left and Its Quarrels with Science* (Baltimore: Johns Hopkins University Press, 1994); Meredith F. Small, "Review of *Primate Visions*, by D. Haraway," *American Journal of Physical Anthropology* 85 (1990): 527–28. See also the responses to the hoax played by physicist Alan Sokal in

Social Text 46–47 (1996). For example, see *Lingua Franca* (July–August 1996): 54–64.

43. The specter of the literary and legal critic Stanley Fish haunts me as I pursue this line of questioning. Although a master of exposing the situatedness of interpretation, he argues against making any connection between becoming "more self-consciously situated [and] inhabit[ing] our situatedness in a more effective way" ("Anti-Foundationalism, Theory Hope, and the Teaching of Composition," in *Doing What Comes Naturally: Change, Rhetoric, and the Practice of Theory in Literary and Legal Studies* [Durham, N.C.: Duke University Press, 1989], 347). He seems to be looking for connections that would have the status of guarantees, the stuff, ironically, of foundationalists. Granted, awareness of the constraints on one's situation does not automatically relax those constraints (ibid., 351), but this has no logical bearing on the empirical and practical question: when and how can systematic reflection on one's situatedness become a resource facilitating reconstruction of one's work? Despite Fish's error in logic, his argument invites attention to his situatedness. At the same time, the popularity of his argument invites attention to the situatedness in the politics of the 1990s of anyone trying to connect politics and analyses of situatedness (see also Downey et al., "Introduction"). (As Bill Lynch reminded me, the popularity during the 1980s of exposing the situatedness of interpretation also invites interpretation, but this is more a matter of historical interest.)
44. Ross, *Strange Weather*, 13.
45. Ibid., 10.
46. Hess, *Science and Technology in a Multicultural World*, chapters 1 and 9.
47. Sharon Traweek, "When Eliza Doolittle Studies 'enry 'iggins," in *Technoscience and Cyberculture*, ed. Aronowitz et al., 37–55.
48. Sharon Traweek, "Tinkering with/in High Energy Physics: International Collaboration in Japan and the United States," lecture given to the Department of Science and Technology Studies, Cornell University, Ithaca, New York, April 26, 1993. See also her *Turbulent Phase Transitions in Japanese and American High Energy Physics* (forthcoming).
49. Haraway, "Mice into Wormholes."
50. Recall also Traweek's Japanese woman high energy physicist in "When Eliza Doolittle Studies." The kin listed here are borrowed and adapted by Haraway respectively from Audre Lorde, Trinh T. Minh-Ha, Native American myths, and Joanna Russ. See Donna J. Haraway, "Manifesto for Cyborgs: Science, Technology, and Socialist feminism in the 1980s," *Socialist Review* 80 (1985): 65–107; "The Promises of Monsters: A Regenerative Politics for Inappropriate/d Others," in *Cultural Studies*, ed. Laurence Grossberg, Cary Nelson, and Paula A. Treichler (New York: Routledge, 1991); "Mice into Wormholes." See also Edwards's essay on cyborgs in this volume and Gray, *Cyborg Handbook*.
51. In "The Promises of Monsters," Haraway speaks of "relentless artifactualism." I read this as a combination of two themes: *(a)* humans (and other organisms, especially primates) are heterogeneous constructors; and *(b)* commodification is inexorable.
52. Haraway, "Situated Knowledges."
53. Or, using a metaphor of Haraway's, to tease out particular strands of the "cat's cradle," but not to tear them out (Donna J. Haraway, "A Game of Cat's Cradle: STS, Feminist Theory, Cultural Studies," *Configurations* 2, no. 1 [1994]: 59–71).
54. Raúl García-Barrios and Luis García-Barrios, "Environmental and Technological Degradation in Peasant Agriculture: A Consequence of Development in Mexico," *World Development* 18, no. 11 (1990): 1569–85. For an overview of political ecology, see Richard Peet and Michael Watts, "Introduction: Development Theory and Environ-

ment in an Age of Market Triumphalism," *Economic Geography* 69, no. 3 (1993): 227–53.

55. Indeed, much remains to be worked out; this image of intersecting processes is not sharply defined. Let me just stress that it is not meant to valorize small interventions; these might be at cross-purposes to other small interventions, and, unless local interveners have complementary visions of the larger intersecting processes, never combine into anything significant.
56. Michael Fortun, "Projecting Speed Genomics," in *The Practices of Human Genetics: International and Interdisciplinary Perspectives* (*Sociology of the Sciences Yearbook*, vol. 19), ed. Michael Fortun and Everett Mendelsohn (Boston: Kluwer, 1996).
57. William Cronon, *Nature's Metropolis: Chicago and the Great West* (New York: Norton, 1991).
58. For allied contributions, see notes 3 and 6 and work cited in the Introduction to this volume.

Contributors

Simon A. Cole, Science and Technology Studies, Cornell University. Cole is developing an anthropological method for analyzing technics. Besides genetic engineering, the research sites for this project are oil exploration and production, security technologies, and DNA fingerprinting.

Paul N. Edwards, Science, Technology, and Society Program, Stanford University. Edwards's cultural history of computers, artificial intelligence, psychology, and the military appeared as *The Closed World* (MIT Press, 1996). He is currently studying the history and politics of climate change research, and directs the Information Technology and Society Project.

Scott F. Gilbert, Biology, Swarthmore College. Gilbert has an M.A. in the history of science, a Ph.D. in biology, and is the author of the textbook *Developmental Biology*. His essay for this collection arose out of lengthy hours on an Educational Policy Committee at Swarthmore.

Herbert Gottweis, Institute for Political Science, University of Salzburg. Gottweis's work in the politics of science analyzes European genetic engineering and biotechnology in the context of developments in political economy and social movements over the last three decades. His book *Governing Molecules: The Discursive Politics of Genetic Engineering in Europe* will be published by MIT Press in 1998.

Yrjö Haila, Regional Studies and Environmental Policy, University of Tampere, Finland. Haila is coauthor of *Humanity and Nature: Ecology, Science, and Society* (Pluto Press) and has written many articles ranging from avian ecology in the taiga to art criticism.

Saul E. Halfon, Science and Technology Studies, Cornell University. Halfon is broadly interested in the politics of science and technology, particularly as evident in policy making at various levels and in local-global connections. His current work focuses on the politics of the overpopulation debates, in an attempt to understand what such debates mean within broader social contexts and how population policy making is sustained over time and distance.

Rosaleen Love, Communication and Language Studies, Victoria University of Technology, Melbourne, Australia. Love writes on issues in Australian science and society in both nonfiction and fiction modes. She has published two collections of short stories with the Women's Press (United Kingdom), *The Total Devotion Machine* (1989), and *Evolution Annie* (1993). She takes particular pleasure in exploring the history of wrong ideas.

Richard A. Schroeder, Geography, Rutgers University. Schroeder spent six years in West Africa as a developer and research fellow with a focus on agrarian and environmental change in Sierra Leone, Nigeria, and The Gambia.

Peter J. Taylor, Eugene Lang Professor of Social Change, Swarthmore College. Focusing on analyses of socioenvironmental complexity since World War II, Taylor's work examines the limitations of and alternatives to the common assumption of systemness and of a global versus a local dichotomy. His political ecology work emphasizes the development of theory that can address intersecting economic, social, and ecological processes operating at different scales. A relevant recent article is "The Social Analysis of Ecological Change: From Systems to Intersecting Processes," *Social Science Information* (1995).

Index